Green Energy and Technology

Climate change, environmental impact and the limited natural resources urge scientific research and novel technical solutions. The monograph series Green Energy and Technology serves as a publishing platform for scientific and technological approaches to "green"—i.e. environmentally friendly and sustainable—technologies. While a focus lies on energy and power supply, it also covers "green" solutions in industrial engineering and engineering design. Green Energy and Technology addresses researchers, advanced students, technical consultants as well as decision makers in industries and politics. Hence, the level of presentation spans from instructional to highly technical.

Indexed in Scopus.

More information about this series at http://www.springer.com/series/8059

Sarah Farrukh · Xianfeng Fan · Kiran Mustafa ·
Arshad Hussain · Muhammad Ayoub ·
Mohammad Younas

Nanotechnology and the Generation of Sustainable Hydrogen

Springer

Sarah Farrukh
School of Chemical and Materials Engineering
National University of Sciences and Technology
Islamabad, Pakistan

Kiran Mustafa
Department of Chemistry
The Women University Multan
Multan, Pakistan

Higher Education Department
Punjab
Pakistan

Muhammad Ayoub
Department of Chemical Engineering
Universiti Teknologi Petronas
Perak, Malaysia

Xianfeng Fan
Department of Chemical Engineering
University of Edinburgh
Edinburgh, UK

Arshad Hussain
Department of Chemical Engineering
National University of Sciences and Technology
Islamabad, Pakistan

Mohammad Younas
Department of Chemical Engineering
University of Engineering and Technology
Peshawar, Pakistan

ISSN 1865-3529 ISSN 1865-3537 (electronic)
Green Energy and Technology
ISBN 978-3-030-60401-1 ISBN 978-3-030-60402-8 (eBook)
https://doi.org/10.1007/978-3-030-60402-8

This Springer imprint is published by the registered company Springer Nature Switzerland AG
The registered company address is: Gewerbestrasse 11, 6330 Cham, Switzerland

I want to dedicate this work to my parents "Abdul Hafeez," "Jamila Hafeez"; my husband "Farrukh Hanif"; and my darling son "M. Sarim Janjua," as without their support and guidance, I am unable to achieve anything in my life.

—Sarah Farrukh

The work is dedicated to my Parents, whose unwavering support is cause of strength; my Brothers, who were always there for me; and my Teacher "Mam Lubna" who first introduced me to the wonderful world of Chemistry.

—Kiran Mustafa.

Acknowledgements

This work is supported by MEMAR lab at School of Chemical and Materials Engineering, NUST, Islamabad and HEC (Pakistan) grant No: 10032/Federal/NRPU/R&D/HEC/2017.

Contents

Chapter 1
Introduction

With the ever-increasing population, the energy demands of the world are continuously increasing. Already existing sources of energy like fossils fuel are harmful for the environment and also depleting day by day. The need of the hour is to look for the sustainable fuels which can satisfy the modern requirements of humans. Hydrogen is the promising alternative to the conventional fuels because of high energy contents, zero greenhouse gas emissions, and low environmental impacts. However, generation and storage of hydrogen involve several difficult procedures to increase the efficiency, and nanotechnology plays an imperative role in increasing the efficiency of these methods. The chapter will explain the various areas of sustainable hydrogen production and storage which employs services of nanotechnology.

1.1 Hydrogen Energy

The report issued by international energy agency states that the energy demands of the world will be twofold by the year 2030 [1]. The consumption of the power varies directly with the human population. The increase in human population resulted in incremental demand of the energy. The population of the world is assumed to be tripled by the year 2030. The most common form of energy is methane or natural gas which includes one carbon atom and four atoms of hydrogen. In order to obtain the maximum of the available resources on earth, the world has started to convert the energy from one form to the other. The fossil fuel energy resources are depleting fast along with being the cause of pollution and global warming. The world is now looking toward renewable energy. Different renewable energy sources like wind energy, solar energy, and waves are promising alternatives to the conventional fossil fuels. However, they are not enough to meet the demands of ever-growing population. From the past few years, hydrogen has emerged as promising fuel and its demand will be skyrocketing in the coming years [2].

S. Farrukh et al., *Nanotechnology and the Generation of Sustainable Hydrogen*,
Green Energy and Technology, https://doi.org/10.1007/978-3-030-60402-8_1

Table 1.1 Energy contents of different fuels

Fuel	Energy content (MJ/Kg)
Hydrogen	120
Liquefied natural gas	54.4
Propane	49.6
Aviation gasoline	46.8
Automotive gasoline	46.4
Automotive diesel	45.6
Ethanol	29.6
Methanol	19.7
Coke	27
Wood (dry)	16.2
Bagasse	9.6

Adapted with permission from ref. [3]. Copyright, Elsevier 2006

Hydrogen is a renewable energy as it is generated from the sources which are inherently renewable such as water. Hydrogen can be converted to the applicable forms of energy such as electricity via the application of small devices like fuel cells. When the hydrogen is converted to electrical energy the by-product is water which is harmless. Other than environment friendly, the energy contents of the hydrogen are also very high. The comparison of energy contents of different fuels is depicted in (Table 1.1).

The hydrogen fuel is important source of clean energy and is significant in overcoming the problem of fossil fuel shortage and has high extension usage. Moreover, the fossil fuels are major cause of harmful pollutants such as oxides of carbon, sulfur, and nitrogen which are produced as the by-product of fossil fuel combustion resulting in the global warming [5]. Different factors of fuels which influence the environment and its efficiency are given by different indicators such as environmental impact factor (EIF), greenization factor (GF), and the content of hydrogen factor (HCF) as shown in Fig. 1.1. The representation of the data in Fig. 1.1 depicts that the hydrogen fuel has lowest impact factor and highest greenization factor, hence making it one of the cleanest fuels.

Hydrogen gas must be produced on large scale and has been researched extensively, in terms of sources for hydrogen generation, methods of hydrogen extraction from various hydrogen sources, processes and techniques of hydrogen storage, and transportation and use of hydrogen in various electronic devices [5]. Sustainable hydrogen can be generated from various different resources, such as water [6], hydrogen sulfide [7], ammonia [8], hydrocarbons [9], biomass [10], etc. Different methods of extraction are involved in hydrogen production from these sources. After hydrogen generation, the next big step is the storage of hydrogen. The H_2 cannot be used as gaseous fuel in various electronic devices as the medium of fuel, it must be stored in a way that can ensure portable applications as well as automotive purposes [11].

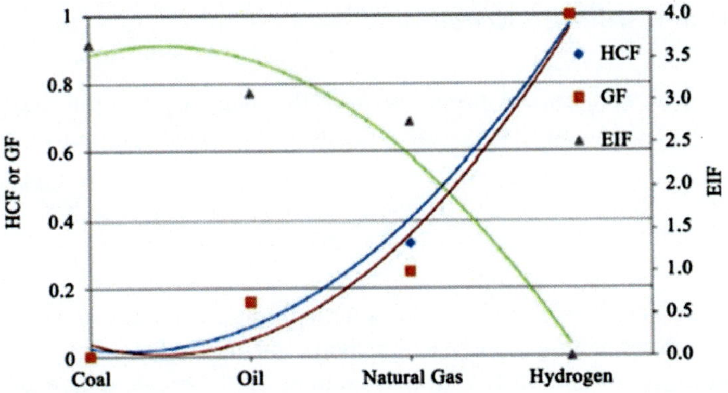

Fig. 1.1 Environmental impact factor (EIF), greenization factor (GF), the content of hydrogen factor (HCF) of different fuels (Adapted with permission from ref. [4] Copyright, Elsevier 2015)

If we look into the past, we can observe that throughout the course of human history, the advances in development of civilizations are deeply related to the discovery and development of novel materials [12]. In several ways, the materials can be regarded as the roots of almost all technologies as many of the technological advances have been obtained through the fabrication of new substances. Nanomaterials are prime example of new material development. Nanotechnology now plays significant role in almost every field of science and technology. In order to obtain the technical advancements and breakthroughs in the area of sustainable hydrogen generation and hydrogen storage, nanomaterials are absolute requirement. Nanomaterials are actively involved in almost every method of generation and storage of hydrogen. The present study will throw light on utilization of nanotechnology and nanomaterials in different methods of hydrogen generation and storage.

1.2 Hydrogen Generation and Nanotechnology

Hydrogen can be generated via several different processes. The sources and processes of hydrogen vary so much that several different ways of classification can be used for their categorization. Here the hydrogen generation is classified on the basis of sources which liberates hydrogen. This work will elucidate the hydrogen generation on the basis of raw material which produces hydrogen. Several methods of hydrogen generation discussed below involve expounded use of nanotechnology one way or another.

1.2.1 *Water Splitting Reaction*

Sustainable hydrogen can be produced from the water by splitting the molecule directly into its elements. The process is known as water splitting reaction. The general reaction is.

$$2H_2O \longrightarrow 2H_2 + O_2$$

The splitting of H-O-H bonds in the water splitting reaction can be achieved by variety of different methods, and each method fundamentally varies from the other in the source of power, as energy is required to break the bonds. There are different power sources involved in the splitting of the molecules like electrical energy, thermal energy or electromagnetic radiation. Normally, mentioned as photolysis, thermolysis, and electrolysis [13].

All of these processes involve the use of nanotechnology for H_2 generation. For instance, Wang et al. achieved photochemical splitting of water with the help of quantum dots. In this investigation, a hybrid structure of carbon quantum dots attached to the single-layer C_3N was studied. The first-principle calculations have depicted that the hybrid is capable of gathering visible and infrared light. These hybrid structures were also able to prevent the mixing of hydrogen and oxygen after their liberation from water spitting. This was due to the fact that the hybrid has sites where redox reaction can occur, and this ensured fast delivery of the photogenerated holes and electrons to the outer C_3N monolayer and inner quantum dots of carbon. The electrostatic forces of attraction forced the protons to penetrate through the monolayer and enter the quantum dots to liberate hydrogen during electrolysis. As no oxygen or hydrogen entered the hybrid, the mixing of the gases was prevented. These metal-free quantum dots-assisted hybrids are appealing candidate for solar energy-driven water splitting reactions with minimum or no mixing of the gases [14].

Similarly, in another study, nanosheets of Ni and rGO were prepared and used for photothermal splitting of water. The hydrogen and oxygen liberation from the water were studied simultaneously by the use of Ni/rGO nanosheets [15]. Nanoparticles of cobalt and nickle alloy have also been reported to possess promising electrocatalytic activity for water splitting reaction. Zhang et al. have reported metal and gas co-doped nanosheets of carbon impregnated with iron-doped Co/Ni alloy. Fe and N have been used for the modification of carbon nanosheets which were later doped with the alloy nanoparticles (FeCoNi@FeNC) in order to achieve better electrocatalytic activity. The obtained material worked as bifunctional catalyst (both for hydrogen and oxygen evolution) in water splitting reaction. The advanced catalytic activity is inherently linked with the existence of Fe in both nanoparticles and nanosheets, as they promote the coordination influence between the nanosheets and nanoparticles. FeCoNi@FeNC is used as both cathode and anode catalyst in water splitting electrolyzer, where it attained the current density of 12 mA cm^{-2} at 1.63 V for 12 h [16].

There are several other studies and investigation which involve the use of nanotechnology for water splitting reaction in order to generate the hydrogen. The detail of such investigations is discussed in Chap. 2.

1.2.2 Decarbonization of Hydrocarbons

Hydrocarbons are also important source of sustainable hydrogen because of their price, availability and convenience in transportation and storage. The carbon and hydrogen are the major constituents of the hydrocarbons, and hence serve as the promising raw material for the generation of hydrogen. The simple chemical reaction of the process can be given as.

$$CH_4 + 2H_2O \rightarrow 4H_2 + CO_2 + \text{energy}$$

Several different methods like steam reforming, partial oxidation, photocatalytic oxidation, and thermal decomposition are used for the decarbonization of hydrocarbons for the production of hydrogen generation [17]. Many of these methods employ nanomaterial for hydrogen liberation from decarbonization of hydrocarbons. For instance, Al_2O_3/NiO nanocatalyst in quartz is used for the steam reforming of methane. The catalyst is prepared by employing sol–gel method. It is found that the size of NiO crystallite in NiO–SiO_2/$Al2O_3$ catalyst varies with the degree of Ni loading and calcination temperature. Al_2O_3/NiO nanocatalyst with 10% Ni is optimized for steam reforming of methane [18]. Similarly, Marin et al. have reported that partial oxidation of Jet-A fuels to hydrogen and carbon monoxide via nanoparticle of MoO_2. These nanoparticles were prepared by the reduction of MoO_3 in 1:3 by volume solution of ethylene glycol and distilled water. The prepared nanoparticle MoO_2 was analyzed with XRD, SEM, BET, and X-ray photoelectron spectroscopy (XPS). MoO_2 nanoparticles depicted ~99% conversion of the fuel (at 850 °C, 1 atm), and ~60% yield of hydrogen. In the similar study, the MoO_2 catalyst was compared with the reference Ni catalyst at similar conditions which was deactivated because of coking in less than 4 h of the process, where MoO_2 nanoparticles showed a 10 h coking resistance [19]. A detailed discussion of nanotechnology and decarbonization of hydrocarbons is provided in Chap. 3.

1.2.3 Decomposition of Hydrogen Sulfide

Hydrogen sulfide is yet another important source of sustainable hydrogen. H_2S is categorized as a highly toxic pollutant of both natural and anthropic origins. Naturally Black Sea can be regarded as the one major reservoir of H_2S because of its internal structure [20]. Coal seams can be considered as the second major natural source of the H_2S [21]. Different anthropic activities are the source of the H_2S generation, such as the extraction processes of petrochemical and oil liberates H_2S in large amounts [22, 23].

General chemical reaction of H_2S conversion to hydrogen can be given by following equations [24]:

$$H_2S + OH^- \leftrightarrow SH - + H_2O$$
$$H_2S + 2OH^- \leftrightarrow S^{2-} + 2H_2O$$
$$2SH^- \rightarrow S_2^{2-} + 2H^+$$
$$2H^+ + 2e^- \rightarrow H_2$$

Different nanomaterials based on transition metal are used in the liberation of hydrogen from hydrogen sulfide, for instance, the nanostructures of $CdIn_2S_4$. A group of investigators synthesized a cubic spinel chemically stable nanostructure of $CdIn_2S_4$ via simple hydrothermal process as visible-light-active photocatalyst for the generation of hydrogen by decomposing H_2S. For the preparation of these nanostructures, mixture of $Cd(NO_3)_2{\cdot}4H_2O$, $In(NO_3)_3{\cdot}3H_2O$ and excessive thiourea was placed in stainless-steel autoclave lined with Teflon along with double distilled water at 140 °C for 60 h. The product was obtained in form of yellow precipitates. The obtained product was given ethanol wash and was dried at 70 °C. The marigold morphology was shown by the samples in aqueous medium samples whereas nanotubes with 25 nm diameter were formed in organic solvent (methanol) under similar conditions. Different characterization techniques like XRD, TEM, and FESEM were used for the analysis of the prepared nanostructures. The nanocatalyst exhibited excellent photocatalytic activity in both mediums. The $CdIn_2S_4$ nanostructures showed give hydrogen quantum yields of 17.1% with nanotubes morphology, whereas 16.8% with marigold-like morphology in visible light. The promising ability of the catalyst for the generation of hydrogen is directly associated with the high crystallinity of the prepared catalyst [25]. More details of nanotechnology and decomposition of hydrogen sulfide are provided in Chap. 4.

1.2.4 Biomass Decomposition

Biomass is an important source for the generation of sustainable hydrogen. Biomass is obtained through variety of sources such as crop residues, waste of agricultural industries, and plants residues like wood and leaves. There are several methods for the generation of H_2 from biomass. These methods can be broadly classified into two categories, i.e., thermochemical processes and biochemical processes [26].

Both the thermochemical and the biological processes of biomass conversion for the generation of sustainable H_2 involve the extensive use of nanotechnology. Many studies have used the nanotechnology for the thermochemical conversion of the algal mass to the sustainable hydrogen [21]. Similarly, various studies of biomass fermentation have reported the incorporation of different nanomaterials with inorganic and organic origins for the production of H_2. For instance, nanomaterials of different metal and metal oxides such as Cu, Au, Pd, Ag, iron-iron-oxide, and TiO_2 are involved in dark fermentation of the biomass for hydrogen generation [22]. Water gas shift reaction is an important biochemical reaction for the generation of hydrogen from biomass. In different studies, various nanomaterials have been utilized in the water gas shift reaction for obtaining high quality and better yield of hydrogen. In a study, researchers have claimed that gold nanoparticles can be used for the generation of hydrogen via water gas shift reaction. The study reported that the method of preparation and the pretreatment of the catalysts are very critical for the functioning of the catalyst in water gas shift reaction. In order to achieve the catalytic activity of the Au nanoparticles in the reaction, the close contact between oxide support and gold is crucial. The close contact is associated with the zerovalent atoms of Au which is necessary for the high catalytic activity. The abovementioned hypothesis is supported by the in situ extended X-ray absorption fine structure (EXAFS) measurements [27]. A detailed discussion of nanotechnology and hydrogen generation from biomass is provided in Chap. 5.

1.2.5 Hydrogen Generation from Miscellaneous Sources

Other than water and fossil fuels, here are several other important substances that generate H2 via chemical or physical treatment. Municipal solid waste is also used as the source of hydrogen. Different components of the waste like biological substances, organic materials, and other undesirable components like plastic debris are involved in the generation of hydrogen. Similarly, several chemical substances like hydrides yield hydrogen via different methods. Variety of the processes involved in hydrogen generation from miscellaneous sources use nanotechnology. For instance, a study has reported the use of nanosized platinum dispersed on $LiCoO_2$ for the generation of hydrogen from the solution of lithium borohydride. The substrate ($LiAlH_4$) liberated a stoichiometric amount of Hydrogen when allowed to react with water, over the prepared 4 nanocatalysts.. The chemical equation for the reaction can be given as.

$$LiBH_4 + 4H_2O \rightarrow LiBO_2 \cdot 2H_2O + 4H_2$$

The said nanocatalysts were prepared by calcining the lithium cobalt powder 250 °C for 5 h, followed by the coating of Pt by employing the oxide of the metal. The said materials were characterized with wide angle X-ray diffraction (XRD), transmission electron microscopy (TEM), and energy-dispersive X-ray spectroscopy [28]. More discussion on nanotechnology-based hydrogen generation from miscellaneous sources is provided in Chap. 6.

1.3 Storage of Hydrogen and Nanotechnology

One of the major reasons of why hydrogen is still not used as the primary source fuel is the limitations in storage of hydrogen. The storage of the hydrogen should be reversible at ambient conditions or it can be released from the storage material on demand. When the hydrogen gas comes in the vicinity of the hydrogen-adsorbing material, some of the approaching molecules (depending upon the nature of the adsorbent) develop physical association or weak van der Waals interaction with the material surface and get physisorbed on it. When sufficiently high energy is provided either by altering the temperature or pressure, the molecules of hydrogen dissociates into atoms and form chemical bonds with the adsorbent, i.e., chemisorption. Reversing the conditions of temperature or pressure releases the adsorbed hydrogen [29].

1.3.1 Physisorption

Physisorption is a non-dissociative surface association of the hydrogen gas with the solid surfaces. The interactions involve weak van der Waals associations between the gaseous hydrogen molecules and the adsorbent. Different nanomaterials like carbons nanotubes (both single walled and multiwalled), zeolites, activated carbons, (COFs), fullerene (nanocages), and covalent-organic frameworks are capable of storing hydrogen by physisorption [3–5]. Different studies have made use of these nanomaterials for physisorption of hydrogen. For instance, an investigation employed first-principle calculation within density functional theory (DFT) for studying the hydrogen storage capability in Jahn–Teller slanted Ti-modified fullerenes. It is observed that Ti atoms make two hexagonal pyramidal structures because of their high cohesive energy. Each Ti atom adsorbed four hydrogen molecules via Kubas interactions, with 0.33–0.76 eV adsorption energy per molecule of hydrogen. The calculations made in the study and the van't Hoff desorption temperature depicted that molecules of hydrogen are reversibly adsorbed under feasible thermodynamic conditions with 10.5 wt. % of hydrogen [13]. Several other studies that use nanotechnology in physisorption of hydrogen are elaborated in Chap. 7.

1.3.2 Chemisorption

Chemical adsorption or chemisorption is another important mechanism of hydrogen storage. In terms of hydrogen storage, any material when gets attached to the hydrogen it becomes a hydride. Over the last few years, the scope of materials to get hydrogenated or hydride formation has immensely expanded. The advancement in the formation of hydrides is accompanied by the fast development in the nanotechnology.

The coupling of the both has led to the remarkable progress in the field of hydrogen storage [1]. Different investigations and analysis have reported the use of nanotechnology for chemisorption of hydrogen. For instance, *ZhiGang* et al. have reported the barrierless physisorption to chemisorption of hydrogen molecules on the fullerenes, doped with lightweight elements. The study employed local density approximation (LDA) in density functional theory (DFT) as process. The DFT process is capable of large systems at low computational costs. In practice, $C_{35}B$ fullerene is brought onto the space of $C_{35}B - H2$ stable system. The two $C_{35}B$ molecules now behave like forceps with each attaching itself to the opposite side of the small H_2 molecules resulting into its dissociation to H, H (with one H attached with each of $C_{35}B$). This is barrierless physisorption to chemisorption. The study suggested that H–H bond breaking occurs because of energy discharge as $C_{35}B$ gets near to the $C_{35}BH2$. The evolved energy increased the bond length between H–H ultimately leading to chemisorption of H_2 molecules in the hydrogen storage method. Overall the energy decreases monotonously during this procedure [11]. Many other investigations that employ nanotechnology in chemisorption of hydrogen are discussed in Chap. 8.

1.3.3 Fuel Cells

Fuel cells are regarded as the green power houses of the modern century. They are thought as small sources with the ability to turn hydrogen energy or economy into reality. The depletion of existing energy sources and increased pollution caused by their production and use are the major reasons behind the immense research in the area of hydrogen fuel cells [1, 2]. A typical hydrogen fuel cell consists of two electrodes (cathode and anode) and an electrolyte membrane (mostly proton exchange membrane, i.e., PEM). The hydrogen and oxygen enter through the anode and cathode of a fuel cell, respectively [6]. The hydrogen fuel cells utilized nanotechnology in variety of ways. In the last few years, fuel cells have exhibited remarkable consistency and lower prices owing to the inclusion of nanomaterials in their production. The involvement of nanotechnology in the fabrication of fuels cells allows high aspect ratio, greater surface area which leads to more power generation, high energy densities, easy miniaturization, and longer shelf life. All of these characteristics are vital for the preparation of powerful fuel cell for transportable electric devices [8, 9]. Some of the fuel cells involve the use of nanomembranes [10] while other uses nanomaterial as electrodes [11]. Nanomaterials are also used as catalysts in some of the fuel cells [12]. A detailed study of nanotechnology and its applications in fuel cells is provided in Chap. 9.

The book aims to discuss in detail the applications of nanotechnology and nanomaterials in various areas of sustainable hydrogen production and storage. The major aim behind the work is to assist the scientists and researchers in the field of sustainable energy.

References

1. Eia US (2017) Monthly energy review
2. Hennicke P, Fischedick M (2006) Towards sustainable energy systems: The related role of hydrogen. Energy Policy 34(11):1260–1270. https://doi.org/10.1016/j.enpol.2005.12.016
3. Ni M, Leung MKH, Sumathy K, Leung DYC (2006) Potential of renewable hydrogen production for energy supply in Hong Kong. Int J Hydrogen Energy 31(10):1401–1412. https://doi.org/10.1016/j.ijhydene.2005.11.005
4. Dincer I, Acar C (2015) Review and evaluation of hydrogen production methods for better sustainability. Int J Hydrogen Energy 40(34):11094–11111. https://doi.org/10.1016/j.ijhydene.2014.12.035
5. Iulianelli A, Basile A (2020) Development of membrane reactor technology for H2 production in reforming process for low-temperature fuel cells. In: Current trends and future developments on (Bio-) Membranes. Elsevier, pp 287–305
6. Lv H, Ruberu TPA, Fleischauer VE, Brennessel WW, Neidig ML, Eisenberg R (2016) Catalytic light-driven generation of hydrogen from water by iron dithiolene complexes. J Am Chem Soc 138(36):11654–11663
7. Guldal NO, Figen HE, Baykara SZ (2017) Production of hydrogen from hydrogen sulfide with perovskite type catalysts: LaMO3. Chem Eng J 313:1354–1363
8. Hayakawa Y, Miura T, Shizuya K, Wakazono S, Tokunaga K, Kambara S (2019) Hydrogen production system combined with a catalytic reactor and a plasma membrane reactor from ammonia. Int J Hydrogen Energy 44(20):9987–9993
9. Jie X, Gonzalez-Cortes S, Xiao T, Yao B, Wang J, Slocombe DR, Fang Y, Miller N, Al-Megren HA, Dilworth JR (2019) The decarbonisation of petroleum and other fossil hydrocarbon fuels for the facile production and safe storage of hydrogen. Energy Environ Sci 12(1):238–249
10. Puga AV (2016) Photocatalytic production of hydrogen from biomass-derived feedstocks. Coord Chem Rev 315:1–66
11. Schlapbach L, Züttel A (2011) Hydrogen-storage materials for mobile applications. In: Materials for sustainable energy: a collection of peer-reviewed research and review articles from nature publishing group. World Scientific, pp 265–270
12. Marquis FDS (2011) The role of nanomaterials systems in energy and environment: renewable energy. JOM 63(1):43
13. Grimes C, Varghese O, Ranjan S (2007) Light, water, hydrogen: the solar generation of hydrogen by water photoelectrolysis. Springer Science and Business Media,
14. Wang X, Jiang X, Sharman E, Yang L, Li X, Zhang G, Zhao J, Luo Y, Jiang J (2019) Isolating hydrogen from oxygen in photocatalytic water splitting with a carbon-quantum-dot/carbon-nitride hybrid. J Mater Chem A 7(11):6143–6148
15. Gu L, Zhang C, Guo Y, Gao J, Yu Y, Zhang B (2019) Enhancing electrocatalytic water splitting activities via photothermal effect over bifunctional nickel/reduced graphene oxide nanosheets. ACS Sustain Chem Eng 7(4):3710–3714
16. Zhang Q, Webster RF, Cheong S, Tilley RD, Lu X, Amal R (2019) Ultrathin Fe-N-C Nanosheets Coordinated Fe-Doped CoNi Alloy Nanoparticles for Electrochemical Water Splitting. Part Part Syst Charact 36(1):1800252. https://doi.org/10.1002/ppsc.201800252
17. Muradov N (2002) Hydrogen from Fossil Fuels without Co2 Emissions. In: Grégoire Padró CE, Lau F (eds) Advances in hydrogen energy. Springer US, Boston, MA, pp 1–16. doi:https://doi.org/10.1007/0-306-46922-7_1
18. Bej B, Pradhan NC, Neogi S (2013) Production of hydrogen by steam reforming of methane over alumina supported nano-NiO/SiO2 catalyst. Catal Today 207:28–35. https://doi.org/10.1016/j.cattod.2012.04.011
19. Marin-Flores O, Turba T, Ellefson C, Scudiero L, Breit J, Norton MG, Ha S (2010) Nanoparticle molybdenum dioxide: a new alternative catalytic material for hydrogen production via partial oxidation of Jet-A fuels. J Nanoelectron Optoelectron 5(2):110–114. https://doi.org/10.1166/jno.2010.1074

20. Naman SA, Veziroğlu A (2013) The low cost hydrogen production from hydrogen sulfide in Black Sea. In: Black sea energy resource development and hydrogen energy problems. Springer, pp 93–108
21. Hoşgörmez H, Yalçın MN, Soylu C, Bahtiyar İ (2014) Origin of the hydrocarbon gases carbon dioxide and hydrogen sulfide in Dodan Field (SE-Turkey). Mar Pet Geol 57:433–444
22. Jarullah AT, Mujtaba IM, Wood AS (2011) Kinetic parameter estimation and simulation of trickle-bed reactor for hydrodesulfurization of crude oil. Chem Eng Sci 66(5):859–871
23. Zuckerman JJ, Hagen AP (1987) Inorganic Reactions and Methods, Vol. 2. The Formation of the Bond to Hydrogen (Part 2). VCH Publishers, Weinheim, Germany,
24. Hawboldt KA, Monnery WD, Svrcek WY (2000) New experimental data and kinetic rate expression for H2S pyrolysis and re-association. Chem Eng Sci 55(5):957–966
25. Kale BB, Baeg JO, Lee SM, Chang H, Moon SJ, Lee CW (2006) CdIn2S4 nanotubes and Marigold nanostructures: a visible-light photocatalyst. Adv Func Mater 16(10):1349–1354
26. Milne TA, Elam CC, Evans RJ (2002) Hydrogen from biomass: state of the art and research challenges. United States. https://doi.org/10.2172/792221
27. Burch R (2006) Gold catalysts for pure hydrogen production in the water–gas shift reaction: activity, structure and reaction mechanism. Phys Chem Chem Phys 8(47):5483–5500
28. Kojima Y, Suzuki K-i, Kawai Y (2006) Hydrogen generation from lithium borohydride solution over nano-sized platinum dispersed on LiCoO2. J Power Sources 155(2):325–328. https://doi.org/10.1016/j.jpowsour.2005.04.019
29. Ma L-L, Rong Y, Han T (2020) A synergistic effect of physisorption and chemisorption: Multiple H2 molecules on the ScC6H6 complex. Int J Hydrogen Energy

Chapter 2
Water Splitting Reactions and Nanotechnology

2.1 Water Splitting Reactions

Sustainable hydrogen can be produced from the water by splitting the molecule directly into its elements. The process is known as water splitting reaction. The general reaction is.

$$2H_2O \longrightarrow 2H_2 + O_2$$

The splitting of H-O-H bonds in the water splitting reaction can be achieved by a variety of different methods, and each method fundamentally varies from the other in the source of power, as energy is required to break the bonds. There are different power sources involving in the splitting of the molecules like electrical energy, thermal energy, or electromagnetic radiation, normally mentioned as electrolysis, thermolysis, and photolysis.

Electrolytic splitting of water is obtained by the conversion of electrical energy to chemical energy as the current is passed through the water. The conversion of energy takes place at the water and electrode interface [1]. The redox reaction occurs, and the hydrogen is obtained at cathode. 3.9% demand of the world hydrogen is fulfilled by this process. The reaction has only oxygen as by-product and "zero" CO_2 emission, and hence it is considered more environmental friendly as compared to the traditional "steam methane reforming" (SMR). Electrochemical cells in which electrolysis of water takes place are run by electricity which is primarily generated by coal or natural gas combustion both of which involves the release of CO_2. Hence, research nowadays is focused on using renewable technologies like photovoltaics or wind turbines for the operation of electrochemical/catalytic water splitting.

Photocatalytic or photochemical splitting of water is a remarkable alternative for the generation of hydrogen. These processes are oriented toward the reduction of CO_2 emission, as they involve the use of renewable sources like sunlight and

S. Farrukh et al., *Nanotechnology and the Generation of Sustainable Hydrogen*, Green Energy and Technology, https://doi.org/10.1007/978-3-030-60402-8_2

water. The electronic bandgap of the photosensitive material must be in alignment with the oxidation and reduction potential of water for photochemical splitting of water [2]. Generally, transition metal cations with d^0 electronic configuration like Ti^{4+}, Ta^{5+}, Nb^{5+}, Zr^{4+}, Ta^{5+}, Mo^{6+}, and W^{6+}, or transition metal cations with a d^{10} electronic configuration like Sn^{4+}, In^{3+}, Ge^{4+}, and Ga^{3+} are found to be significant for photocatalytic substances. The vacant sp- or d-orbitals develops the base of their particular conduction bands [3].

Thermochemical or thermal splitting of water is another technology for the generation of hydrogen with little or no greenhouse gas release. This technique has been widely explored by the scientist and about 300 plus water splitting sequences along with their features, applications, and limitations have been reported. Thermochemical splitting of water needs high temperatures ranging from 500 °C to 2000 °C for the water splitting chemical cycles. The chemicals employed in the methods are used again and again within the cycle, hence forming a loop that uses water only and generates oxygen and hydrogen. This technology uses heat from existing nuclear power stations pathway or concentrated energy from heliostats (solar power) for splitting of water [4]. These are some of the methods by which hydrogen is produced by the splitting of water.

Many of these processes involve nanotechnology at one point or another. In fact, the nanomaterials are frequently employed as the catalysts and electrodes in water splitting reactions. For instance, thermal decomposition of methane. Monometallic nanocatalysts like NiO, Co_3O_4, ZrO_2, and ZnO are used for the purpose. These catalysts are synthesized by a solution combustion process and are characterized by different techniques like Brunauer–Emmett–Teller (BET), X-ray diffraction (XRD), and scanning electron microscopy (SEM). These nanomaterials are able to liberate hydrogen from methane at 850 °C temperature because of their large surface area, highly electropositive nature, fine-tuning of bandgap, and thermal stability. Nanocubanes are employed for the effective splitting of the water and nonporous materials are also employed for the purpose. Photoelectrocatalytic conversion of solar energy also liberates sustainable hydrogen via application of nanomaterials. Photoelectrochemical (PEC) splitting of water employs p- and n-type semiconductors. In a study, these two different photoelectrodes were joined together and hence the redox reactions can occur simultaneously for the efficient use of light. In order to enhance the effectiveness of the PEC, recombination of the photogenerated carrier must be minimized. If nanomaterials are employed then the charge carriers are produced at their surface due to their shape, enhanced surface-to-volume ratio, and precise morphology and hence the splitting of water will take place at the surface. It is observed that the efficiency of PEC increased by 50–90% with the use of nanomaterials. Similarly, several other processes are there which employ nanotechnology for the production of sustainable hydrogen [5]. The present chapter explains the utilization of different nanomaterials like nanotubes, metallic nanoparticles, nanocubanes, etc. in water splitting reaction.

2.2 Quantum Dots and Water Splitting Reaction

Quantum dots are very significant nanoentities. They are nanoscale crystals with the ability to transport electrons. They have widespread applications in water splitting reactions. In a study conducted by Wang et al. a useful set of quantum dots were discovered which have proven their worth as promising electrocatalysts for water splitting reaction and liberation of hydrogen. In this investigation, a hybrid structure of carbon quantum dots attached to the single-layer C_3N was studied. The first-principle calculations have depicted that the hybrid is capable of gathering visible and infrared light. These hybrid structures were also able to prevent the mixing of hydrogen and oxygen after their liberation from water splitting. This was due to the fact that the hybrid has sites where redox reaction can occur, and this ensured fast deliveries of the photogenerated holes and electrons to the outer C_3N monolayer and inner quantum dots of carbon. The electrostatic forces of attraction forced the protons to penetrate through the monolayer and enter the quantum dots to liberate hydrogen during electrolysis. As no oxygen or hydrogen entered the hybrid, the mixing of the gases was prevented. These metal-free quantum dots-assisted hybrids are appealing candidate for solar energy-driven water splitting reactions with minimum or no mixing of the gases [6].

Quantum dots prepared with copper oxide and supported on titanium oxide nanosheets are reported to have excellent hydrogen liberation ability via water splitting reaction. The hybrid of CuO quantum dots and TiO_2 has shown the hydrogen liberation rate of ~0.04 mmol h^{-1} which is approximately 20 times greater than pure nanosheets of TiO_2. This CuO/TiO_2 hybrid was prepared by mixing of titanium (IV) butoxide and HCl. This mixture after stirring for 30 min was placed in Teflon autoclave where hydrothermal reaction occurs, and precipitates were formed which were separated by centrifugation. The residual fluoride ions were removed by treatment of the product with NaOH, and afterward the product was rinsed with deionized (DI) water and fine powder of TiO_2 was obtained. Later on, CuO quantum dots were deposited on the prepared nanosheets by the general hydrothermal procedure. In this process, $CuCl_2$ was added in the ethanol suspension of TiO_2 nanosheets. After continuous stirring for 5 h, ethanol was evaporated in drying oven. Later on, the product was calcinated at 400 °C in muffle furnace, as the product cooled the CuO quantum dots accumulated on the TiO_2 nanosheets. The prepared product was characterized with XRD, XPS, SEM, TEM, and UV–vis diffusion reflectance spectra (DRS) [7].

Ye et al. have reported Pt quantum dot ornamented αFe_2O_3 nanosheets for efficient liberation of hydrogen gas from electrochemical splitting of water. These nanostructures were found to be very efficient in water splitting reaction as they have several active sites, and they exhibited low overpotential i-e 90 mV@10 mA cm^{-2} for hydrogen evolution reaction and a considerably low voltage of 1.51 V in alkaline electrolyte. The flower-like nanostructure was prepared by the electrodeposition and impregnation deposition. The pretreated Ni-foam and graphite were used as cathode and anode, and the mixture of $FeSO_4{\cdot}7H_2O$, Na_2SO_4, and cetyl trimethylammonium bromide (CTAB) was used as the electrolyte. By passing the 10 mA current through the electrochemical cell, αFe_2O_3/Ni electrode was obtained. Afterward, it was immersed in the solution of H_2PtCl_6 for impregnation deposition and was later soaked in $NaBH_4$ for the reduction of Pt. The product was washed with deionized (DI) water. These flower-like Pt quantum dot ornamented αFe_2O_3 nanosheets were characterized with TEM, XRD, field-emission scanning electron microscope (FESEM), and X-ray photoelectron spectroscopy (XPS) [8].

Graphene quantum dots also have favorable electrocatalytic effects in a water splitting reaction. In an investigation, a group of researchers have shown the water splitting abilities of graphene hydrogel/B-doped graphene quantum dots (GH-BDGQD). These quantum dots have high porosity, large surface area, numerous active sites, improved ion diffusion, and mass transport. Because of this the GH-BDGQD exhibits advanced trifunctional electrocatalytic ability including hydrogen liberation, oxygen liberation, and oxygen reduction along with the greater stability as compared to the available commercial counterparts [9].

Z-scheme system of black and red phosphorous quantum dots is capable of achieving water splitting reaction for evolution of hydrogen in the absence of sacrificial agents. In a study, wet-chemistry path was followed for the fabrication of Z-scheme system of black/red phosphorus quantum dots (BRPQD). These quantum dots were fabricated by the conversion of red phosphorous into black phosphorus with reflux reaction between red phosphorous and ethylenediamine. After which phase transformation reaction was allowed to occur and the product was centrifuged at 6000 rpm in order to remove large particles. Later on, Tyndall effect of the particle was observed which indicated fabrication of the stable quantum dots. The synthesized product was characterized with TEM, XPS, UV–Vis spectrometer, UV photoelectron spectroscopy (UPS), time-resolved transient absorption spectroscopy (TAS), and inductively coupled plasma atomic emission spectroscopy (ICP-AES) [10].

A successful research has suggested solenoid nitridation reduction process for the generation of α-MoC_{1-x} quantum dots encased in carbon doped with nitrogen (α-MCNCQD). The frog egg-shaped quantum dots were to promote H_2 evolution in water splitting reaction at all pH levels, with the overpotential of 118 mV, and 10 mA cm^{-2} was obtained in alkaline solutions. The α-MCNCQD also depicted rapidly increasing current density (which was much greater than the commercially available Pt/C electrocatalysts). These quantum dots showed equally promising results in neutral and acidic solutions. α-MCNCQD were prepared by mixing of urea and ammonium molybdate tetrahydrate under the temperature-controlled reduction. Nitrogen atmosphere was provided during the thermal process, which helps in the achievement of N-doping of the mixture; afterward, the product was sonicated and α-MCNCQD were achieved. The prepared quantum dots were characterized with XRD, XPS, FESEM, and TEM and the composition of the catalyst was determined with ICP-AES [11].

2.3 Nanoparticles and Water Splitting Reactions

Nanoparticles have been the integral part of water splitting reaction for evolution of H_2 since the introduction of the reaction for gas liberation. Several studies have reported the use of nanoparticles for the generation of H_2 by splitting of water. Ni and Ru core–shell nanoparticles ornamented on carbon nanosheets (Ni@Ru/CNS-x) have depicted superior photocatalytic activity in splitting of water for liberation of H_2 at all pH levels. Ni@Ru/CNS-x were prepared by facile by using metal organic framework (MOF) as a sacrificial precursor wet-chemistry technique. The Ni@Ru/CNS-x nanoparticles were characterized by SEM and TEM. The prepared catalysts exhibited 10 mA cm^{-2} HER for low overpotential of 20.1 mV in alkaline environment which is far superior than the commercially available Pt/C and RuO_2 [12].

In an investigation conducted by Yu et al. exo-solution of Ag nanoparticles on $AgTaO_3$-$SrTiO_3$ is reported as remarkable plasmonic photocatalysts for splitting of water. Polymers of $AgTaO_3$ and $SrTiO_3$ salts were prepared by standard polymerized complex method. The brown resin obtained was grounded to powder and pressed into pellets which were calcined in muffle furnace, and the material was powdered again. Afterward, the Ag nanoparticles were grown on the freshly prepared powders by making the suspension of the solid powder in ethylene glycol. For the termination of the reaction, the mixture was transferred to the ice water. The product thus obtained was centrifuged several times and rinsed with DI water and dried for 24 h. The prepared material was characterized with XRD, TEM, XPS, and UV–Vis spectrophotometer. Photoelectrochemical analysis of these nanoparticles has shown improved photocatalytic activity and better charge separating conditions [13].

In a recent study, a super-fast synthesis technique of CoS nanoparticles has been reported for the HER by splitting of water. Chen et al. have reported ~7 ms high-temperature treatment method chalcogenide nanoparticles. For the preparation of these nanoparticles, cobalt acetate and thiourea were dissolved in DI water. Later on, graphene oxide ink was added to the solution and the mixture was sonicated and was casted on the glass plate. The graphene sheet was removed after an hour and was annealed in Ar atmosphere and was later on subjected to the thermal shock of about 2000 K which instantly cut the film into small particles. The prepared nanoparticles were characterized with SEM, TEM, XPS, and XRD. CoS and graphene core–shell nanoparticles depict excellent water splitting activity of 10 mA cm^{-2} at a low overpotential of ~1.77 V without any loss in the activity for continuous 60-h operation [14].

2.4 Nanotubes and Water Splitting Reactions

Nanotubes of different materials are also proven to be very efficient in a variety of water splitting mechanisms. For instance, crystalline nanotube of TiO_2. It is one of the most classic materials to improve the efficiency of the water splitting reactions due to their chemical inertness, photostability, and low cost. In a study, it is found that crystalline nanotube of TiO_2 shows increased generation of hydrogen by PEC water splitting as compared to the amorphous TiO_2 nanotube. However, the amorphous nanotubes can be converted to the crystalline form at high temperatures like 300 °C. Highly ordered TiO_2 nanotubes show increased hydrogen generation by PEC water splitting. In order to obtain these nanotubes, the anodization of 0.1-mm-thick titanium foils was performed in ethylene glycol containing NH_4F, H_2O in the presence of Pt counterelectrode, and 50 V potential difference. The experiment took 5 h. Afterward, the newly formed nanomaterials were ultrasonicated with acetone so as to separate the anodized TiO_2 nanotubular from the titanium substrate. After rinsing with DI water, second anodization was conducted on the titanium substrate at the same temperature for the additional growth of nanotubes and the rest of the procedure was repeated. After repeating the whole process again, the obtained nanotubes were dried in the stream of nitrogen. The annealing of the nanotubes was carried out at 300–600 °C. The nanotubes were characterized SEM and XRD [15].

Niobium-doped TiO_2 has also been reported to aid the water splitting reaction by PEC. These doped nanotubes are obtained by growing them on the Nb-Ti alloy. These doped nanotubes depict 2.5 times higher current as compared to the undoped TiO_2 nanotubes [16]. In another study, carbon-doped nanotubes have been developed. These nanotubes depict 20 times more higher photocurrent densities, as compared to the undoped TiO_2 under visible light illumination [17].

In nickel–cobalt bimetal phosphides, nanotubes are also found important for solar-driven water splitting reactions. These nanotubes were prepared by two steps solid-state reaction, where first being the oxidation reaction occurring at 350 °C in normal air atmosphere (for the development of Co_xNi_yO) from metal organic framework (MOF)-74 cobalt–nickel substrate and the second reaction involves the phosphorization and calcination at 300 °C with NaH_2PO_2 in a pure N_2 atmosphere to afford (for the development of Co_xNi_yP). Distinctive phases of bimetal phosphides were attained from altering the Co/Ni ratios in the precursor of MOF. SEM was used to determine the phosphorization of the MOF-74. It was observed that despite phosphorization and calcination the MOF retained its original nanorod morphology. However, the images showed that the modified MOF has the rough surface as compared to the original substrate. From the transmission electron microscope (TEM) images of Co_4Ni_1P, a tubular arcade was found running across the nanorods forming morphology was observed that, usual nanotube structure. The Co_4Ni_1P nanotubes have depicted remarkable hydrogen emission reaction (HER) and oxygen emission reaction (OER) through water splitting. In alkaline solution, the catalyst was found to achieve 10 m A cm^{-2} current density at a voltage of 1.59 V for both electrodes [18].

Non-metallic nanotubes are also proven to improve efficiency of HER from water splitting. In a study N, S-co-doped carbon nanotubes (CNT) were prepared in situ by using polydopamine. The prepared N, S-CNT were simultaneously applied as the electrocatalysts for HER and OER. These nanotubes depicted excellent activity both for the HER and OER from splitting of water. For the fabrication of N, S-CNT, dopamine (DA) was mixed with oxidized CNT dispersion and afterward dissolved into the water and subjected to sonication. Then, phosphate buffered saline was added at room temperature with continued stirring for a day. Later on, 2-mercaptoethanol was added to the mixture and was stirred for half a day. The carbonization of this mixture in N_2 atmosphere resulted in the formation of N, S-CNT [19].

A study has determined that p-type Zn-doped α-Fe_2O_3 nanotube formed on a transparent conductive has excellent photoelectrochemical abilities and assists the overall water splitting reaction. For developing these nanotubes, ZnO nanorod arrays on FTO-coated glass sacrificial templates and Zn sources were used, and $FeCl_2$ aqueous solution was employed as the electrolyte. Afterward, anodic potential of 1 V was applied on the ZnO nanorods for so as to assist the oxidation of ferrous to ferric subsequently leading to the precipitation of the Fe^{3+} as amorphous γ-FeOOH on ZnO nanorod surface. Some of the dissolved Zn ions were deposited on γ-FeOOH and were later on annealed for 120 min in N_2 atmosphere in order to convert γ-FeOOH into α-Fe_2O_3 nanotubes [20].

In an investigation, 1D-1D metal-free multiwalled carbon nanotubes (MWCNT)/SiC nanowires were fabricated via an in situ chemical mechanism between silicon powder and MWCNTs. These nanoheterostructures were grown via vapor–liquid–solid (VLS) mechanism. The (MWCNT)/SiC nanotubes were characterized for their structure, composition, and morphology by XRD, TEM, thermal gravimetric analysis (TGA), and UV–Vis analysis. In the study, the photoactivities of the prepared nanostructures were also determined for the evolution of H_2. The results depicted better activity of metal-free MWCNTs/SiC 1D-1D nanoheterostructures as compared to the simple SiC nanowires in visible light irradiation. The high-efficiency MWCNTs/SiC in liberation of H_2 from water splitting is attributed to its promising separation of photogenerated electron–hole pair, improved visible light absorption, enhanced crystallinity, and distinctive 1D-1D nanoheterostructures [21].

A study reported camphor sulfonic acid doped polyaniline-WO_3 (CSPA-WO_3) nanocomposites modified with reduced graphene oxide (rGO). The CSPA-WO_3 was able to reduce GO to rGO photoelectrochemically due to its semi-conductive nature. The morphology, chemical composition, and structure of CSPA-WO_3-rGO were characterized by X-ray photoelectron spectroscopy (XPS), XRD, Fourier transform infrared (FTIR) spectroscopy, Raman spectroscopy, field-emission scanning electron microscopy (FESEM), and TEM. The influence of rGO addition on the solar. CSPA-WO_3 depicted remarkable photoelectrochemical current density and photoconversion efficiency visible light illumination. For the fabrication of these nanostructures, a suspension of WO_3 nanoparticles and camphor sulfonic acid was made in distilled water. Afterward, distilled aniline was added at 4 °C with continuous stirring. Later on, an oxidizing agent (ammonium persulfate) was slowly added and the prepared CSPA-WO_3 was filtered and washed with water. GO was prepared with modified Hummer's method. For the preparation of CSPA-WO_3-rGO, CSPA-WO_3 and rGO were sonicated separately in water afterward both suspensions were stirred together in a homogenized mixture and were later irradiated by a 300 W Xenon for 60 min while stirring slowly. The material thus obtained was washed and dried at 60 °C [22].

2.5 Nanosheets and Water Splitting Reactions

Nanosheets are also proven to be very effective in water splitting reactions. In a study conducted by Zhou. et al. SiC ultrathin nanosheets covered with GO have been reported. The sheets were prepared by vapor–solid mechanism. These nanosheets have shown promising photocatalytic activities as assessed by HER from pure water and water containing Na2S.SiC/GO nanosheets showed 10 times better photocatalytic activity in H_2 generation as compared to the SiC nanocrystals. The improved photocatalytic of the nanosheets is related to the two-dimensional structure of nanosheet, large surface area, improved visible light absorption, and fast SiC-to-graphene interfacial charge transfer. For the fabrication of these sheets, graphene sheets were mixed with Si powder and were placed inside the tube furnace and heated up to 1320 °C in Ar atmosphere. In the end, the excess Si was removed by treating them with the mixture of HNO_3 and HF and dried at 100 °C. The nanosheets were characterized by XRD, UV–Vis spectroscopy, Raman spectroscopy, TEM, etc. [23].

Semiconductors of tantalate are promising photocatalysts for liberation of hydrogen by photocatalytic splitting of water, as their conduction band involves 5d orbital, which has greater negative as compared to the $H+/H_2$ half reaction. Bi_3TaO_7 is a stable tantalate, having appropriate bandgap for visible light illumination under acidic or alkaline conditions. Nanosheets made of Bi_3TaO_7 are found to be significant for hydrogen generation from splitting of water. They are prepared by mixing the ethanol suspensions of Bi $(NO_3)_3{\cdot}5H_2O$ and $TaCl_5$. Afterward, the pH of the solution was adjusted aqueous solution of KOH. Later on, the mixture was placed in Teflon-lined stainless-steel autoclave for a day at 180 °C. In the end, the Bi_3TaO_7 nanoparticles were obtained, they were cooled at room temperature and washed with water. Bi_3TaO_7 nanosheet was generated by the liquid exfoliation process. For this purpose, suspension of $Bi(NO_3)_3 \cdot 5H_2O$ nanoparticles was made in distilled water having sodium dodecyl sulfate (SDS) as the surfactant. The non-polar hydrocarbon tail of the SDS combined with the surface of the nanosheet and the polar head of SDS combined with the water lead to the formation of steric repulsion-resistant nanosheets. Later on, the sheets were sonicated and centrifuged. These nanosheets were characterized with XRD, Raman spectroscopy, SEM, and EDS. Nanoparticles of Bi_3TaO_7 depicted no hydrogen liberation; however, Bi_3TaO_7 nanosheets showed ~2.1 μ mol/h liberation of hydrogen in water splitting reaction [24].

In a study, nanosheets of Ni and rGO were prepared for photothermal effect-driven water splitting reaction. The hydrogen and oxygen liberation from the water was studied simultaneously by the use of Ni/rGO nanosheets. Over the nanosheets of Ni/rGO, under the illumination of visible light, both hydrogen and oxygen liberations were increased promisingly as the overpotential reduced to 49 and 50 mV at 10 mA cm^{-2}, respectively. The rGO absorbed incident irradiation for improving the conductivity of electrocatalyst and heating the supported active entity (i.e Ni). The hot active entity can enable the thermodynamics and kinetics of electrocatalytic reactions and hence highly improve electrocatalytic performance. For the preparation of these nanosheets Ni $(OH)_2$ nanoparticles were prepared first over the surface of rGO by hydrothermal process. Later on, the Ni/rGO composite was made by the reduction of $Ni(OH)_2$/rGO at 500 °C in Ar/H_2 atmosphere. These nanosheets were characterized by infrared (IR) imaging, TEM, XRD, Raman spectroscopy, and X-ray photoelectron spectroscopy (XPS) [25].

In a relatively new research, cauliflower-like nanosheets have been developed with superior photocatalytic activity for generation of hydrogen with water splitting reaction. Ternary ZnS/CuS/g-C_3N_4 cauliflower-like composite nanosheets were fabricated via hydrothermal, cation exchange process along with ultrasound supported wet impregnation techniques. These nanosheets have proven to be very efficient for generation of hydrogen. Maximum hydrogen liberation rate of 9868 μ mol h^{-1} g^{-1} was obtained. The hydrogen liberated by ternary ZnS/CuS/g-C_3N_4 is almost double than the amount generated by simple CuS/ZnS. This is maximum amount of hydrogen generated by g-C_3N_4-based nanocomposites. This remarkable performance is attributed to the exclusive heterostructure of ternary CuS/ZnS/g-C_3N_4 which have possibly restrained the rejoining of electron/hole pairs. The nanosheets also showed remarkable recycling performance. For the preparation of ternary ZnS/CuS/g-C_3N_4, ZnS, CuS/ZnS nanocomposites, and g-C_3N_4 were prepared separately by hydrothermal method, cation exchange method, and heating the melamine, respectively. Afterward, the nanosheets were fabricated by ultrasonic-aided wet impregnation technique. The prepared nanostructures were characterized by FTIR, XRD, XPS, SEM, TEM, etc. [26].

Snowflake-like multi-channel Ru/Cu nanosheets are also reported for their efficiency as electrocatalysts for HER and OER via water splitting reactions. These metallic nanosheets were composed of amorphous Cu and crystalline Ru which depicted efficient liberation of hydrogen and oxygen from water splitting in both alkaline and acidic conditions at lower onset and overpotential as compared to the commercially available $\mathrm{Ir}/\mathrm{C}\|\mathrm{Pt}/\mathrm{C}$ Ir/C||Pt/C electrocatalyst. These nanosheets have also shown excellent stability in both alkaline and acidic environments as indicated by the chronoamperometry measurements. These nanosheets were prepared by the one-pot process with $RuCl_3 \cdot xH2O$ and $CuCl_2 \cdot 2H_2O$. Oleylamine and octadecene were employed as solvents and phloroglucinol as the reducing agent. The prepared nanosheets were characterized by TEM, AFM, EDX, dark-field scanning TEM, etc. [27].

2.6 Other Nanomaterials and Water Splitting Reaction

Other than abovementioned nanomaterials there are several other nanoscale substances that are involved in the water splitting reaction for the generation of hydrogen. Some of them are listed in Table 1.1.

Table 1.1 Water splitting reaction for the generation of hydrogen and nanotechnology

Material	Nanotechnology	Function	References
$BaTiO_3$	Ferroelectric nanoparticles	Ultrasonic vibrations for water splitting	[28]
$SrTiO_3$	Nanocubes	Electrochemical water splitting	[29]
$SmMn_2O_5$	Nanocuboids	Electrochemical water splitting	[30]
Cobalt phosphide	Nanopolyhedrons	Electrochemical water splitting	[31]
Ni/NiCoP	Nanoheterojunction	Electrocatalytic water splitting	[32]
Brass	Nanopatterning	Photoelectrochemical water splitting	[33]
GaN	Nanowires	Photoelectrochemical water splitting	[34]
CoS_2–MoS_2	Nanoflakes	Electrochemical water splitting	[35]
$Mo_xW_{1\text{-}x}O_2/Mo_xW_{1\text{-}x}S_2/Mo_xW_{1\text{-}x}P$	Nanooctahedrons	Electrocatalytic water splitting	[36]
$MoS_2/Co_9S_8/Ni_3S_2/Ni$	Nanoassembly	Electrocatalytic water splitting	[37]

2.7 Conclusion

The above study has explained the use of various nanomaterials in water splitting reaction for the generation of hydrogen. A wide variety of nanomaterials including the normal metal nanoparticles, the transition metal-based nanomaterials, the carbon-based nanostructures, etc. are found to improve the efficiency of the water splitting reactions. In some cases, the nanomaterials have found to increase the hydrogen liberation rate by 20-folds as compared to the bulk material (the quantum dots of CuO/TiO_2 hybrid). The study concludes that the nanomaterials are integral for future endeavors in the area of water splitting for hydrogen generation. The need of the hour is to look for the processes which enable the economical synthesis of the nanomaterials and also the alternatives for expensive raw materials (such as noble metals) for the synthesis of nanomaterials must be researched.

References

1. Grimes C, Varghese O, Ranjan S (2007) Light, water, hydrogen: the solar generation of hydrogen by water photoelectrolysis. Springer Science and Business Media
2. Kochuveedu ST (2016) Photocatalytic and photoelectrochemical water splitting on TiO2 via photosensitization. J Nanomater
3. Kadassery KJ, Dey SK, Cannella AF, Surendhran R, Lacy DC (2017) Photochemical water-splitting with organomanganese complexes. Inorg Chem 56(16):9954–9965
4. Vozniuk O, Tanchoux N, Millet J-M, Albonetti S, Di Renzo F, Cavani F (2019) Spinel mixed oxides for chemical-loop reforming: from solid state to potential application. In: Studies in surface science and catalysis, vol 178. Elsevier, pp 281–302
5. Szymanski P, El-Sayed MA (2014) Some recent developments in photoelectrochemical water splitting using nanostructured TiO 2: a short review. In: Marco Antonio Chaer Nascimento. Springer, pp 7–18
6. Wang X, Jiang X, Sharman E, Yang L, Li X, Zhang G, Zhao J, Luo Y, Jiang J (2019) Isolating hydrogen from oxygen in photocatalytic water splitting with a carbon-quantum-dot/carbon-nitride hybrid. J Mater Chem A 7(11):6143–6148
7. Wang Y, Zhou M, He Y, Zhou Z, Sun Z (2020) In situ loading CuO quantum dots on TiO2 nanosheets as cocatalyst for improved photocatalytic water splitting. J Alloy Compd 813:152184
8. Ye B, Huang L, Hou Y, Jiang R, Sun L, Yu Z, Zhang B, Huang Y, Zhang Y (2019) Pt (111) quantum dot decorated flower-like αFe 2 O 3 (104) thin film nanosheets as a highly efficient bifunctional electrocatalyst for overall water splitting. J Mater Chem A 7(18):11379–11386
9. Tam TV, Kang SG, Kim MH, Lee SG, Hur SH, Chung JS, Choi WM (2019) Novel graphene hydrogel/B-doped graphene quantum dots composites as trifunctional electrocatalysts for Zn–air batteries and overall water splitting. Adv Energy Mater 9(26):1900945
10. Shi R, Liu F, Wang Z, Weng Y, Chen Y (2019) Black/red phosphorus quantum dots for photocatalytic water splitting: from a type I heterostructure to a Z-scheme system. Chem Commun 55(83):12531–12534
11. Lin L, Sun Z, Yuan M, Yang H, Li H, Nan C, Jiang H, Ge S, Sun G (2019) α-MoC1–x Quantum dots encapsulated in nitrogen-doped carbon for hydrogen evolution reaction at All pH values. ACS Sustain Chem Eng 7(10):9637–9645
12. Wu W, Wu Y, Zheng D, Wang K, Tang Z (2019) Ni@ Ru core-shell nanoparticles on flower-like carbon nanosheets for hydrogen evolution reaction at All-pH values, oxygen evolution reaction and overall water splitting in alkaline solution. Electrochim Acta 320:134568
13. Yu J, Zhang L, Qian J, Zhu Z, Ni S, Liu G, Xu X (2019) In situ exsolution of silver nanoparticles on AgTaO3-SrTiO3 solid solutions as efficient plasmonic photocatalysts for water splitting. Appl Catal B 256:117818
14. Chen Y, Xu S, Zhu S, Jacob RJ, Pastel G, Wang Y, Li Y, Dai J, Chen F, Xie H (2019) Millisecond synthesis of CoS nanoparticles for highly efficient overall water splitting. Nano Res 12(9):2259–2267
15. Gong J, Lai Y, Lin C (2010) Electrochemically multi-anodized TiO2 nanotube arrays for enhancing hydrogen generation by photoelectrocatalytic water splitting. Electrochim Acta 55(16):4776–4782
16. Das C, Roy P, Yang M, Jha H, Schmuki P (2011) Nb doped TiO 2 nanotubes for enhanced photoelectrochemical water-splitting. Nanoscale 3(8):3094–3096
17. Park JH, Kim S, Bard AJ (2006) Novel carbon-doped TiO2 nanotube arrays with high aspect ratios for efficient solar water splitting. Nano Lett 6(1):24–28
18. Yan L, Cao L, Dai P, Gu X, Liu D, Li L, Wang Y, Zhao X (2017) Metal-Organic Frameworks Derived Nanotube of Nickel-Cobalt Bimetal Phosphides as Highly Efficient Electrocatalysts for Overall Water Splitting. Adv Func Mater 27(40):1703455
19. Qu K, Zheng Y, Jiao Y, Zhang X, Dai S, Qiao SZ (2017) Polydopamine-Inspired, Dual Heteroatom-Doped Carbon Nanotubes for Highly Efficient Overall Water Splitting. Adv Energy Mater 7(9):1602068

20. Qi X, She G, Wang M, Mu L, Shi W (2013) Electrochemical synthesis of p-type Zn-doped α-Fe 2 O 3 nanotube arrays for photoelectrochemical water splitting. Chem Commun 49(51):5742–5744
21. Zhou X, Li X, Gao Q, Yuan J, Wen J, Fang Y, Liu W, Zhang S, Liu Y (2015) Metal-free carbon nanotube–SiC nanowire heterostructures with enhanced photocatalytic H 2 evolution under visible light irradiation. Catal Sci Technol 5(5):2798–2806
22. Hosseini MG, Sefidi PY, Aydin Z, Kinayyigit S (2020) Toward enhancing the photoelectrochemical water splitting efficiency of organic acid doped polyaniline-WO3 photoanode by photo-assisted electrochemically reduced graphene oxide. Electrochim Acta 333:135475
23. Zhou X, Gao Q, Li X, Liu Y, Zhang S, Fang Y, Li J (2015) Ultra-thin SiC layer covered graphene nanosheets as advanced photocatalysts for hydrogen evolution. J Mater Chem A 3(20):10999–11005
24. Razavi-Khosroshahi H, Mohammadzadeh S, Hojamberdiev M, Kitano S, Yamauchi M, Fuji M (2019) Visible light active Bi 3 TaO 7 nanosheets for water splitting. Dalton Trans 48(25):9284–9290
25. Gu L, Zhang C, Guo Y, Gao J, Yu Y, Zhang B (2019) Enhancing electrocatalytic water splitting activities via photothermal effect over bifunctional nickel/reduced graphene oxide nanosheets. ACS Sustain Chem Eng 7(4):3710–3714
26. Rameshbabu R, Ravi P, Sathish M (2019) Cauliflower-like CuS/ZnS nanocomposites decorated g-C3N4 nanosheets as noble metal-free photocatalyst for superior photocatalytic water splitting. Chem Eng J 360:1277–1286
27. Yao Q, Huang B, Zhang N, Sun M, Shao Q, Huang X (2019) Channel-Rich RuCu Nanosheets for pH-Universal Overall Water Splitting Electrocatalysis. Angew Chem Int Ed 58(39):13983–13988
28. Su R, Hsain HA, Wu M, Zhang D, Hu X, Wang Z, Wang X, Ft Li, Chen X, Zhu L (2019) Nanoferroelectric for high efficiency overall water splitting under ultrasonic vibration. Angew Chem Int Ed 58(42):15076–15081
29. Klusáčková M, Nebel R, Macounová KM, Klementova M, Krtil P (2019) Size control of the photo-electrochemical water splitting activity of SrTiO3 nano-cubes. Electrochim Acta 297:215–222
30. Rani BJ, Ravi G, Yuvakkumar R, Hong SI (2019) Novel SmMn2O5 hollow long nano-cuboids for electrochemical supercapacitor and water splitting applications. Vacuum 166:279–285
31. Li J-S, Kong L-X, Wu Z, Zhang S, Yang X-Y, Sha J-Q, Liu G-D (2019) Polydopamine-assisted construction of cobalt phosphide encapsulated in N-doped carbon porous polyhedrons for enhanced overall water splitting. Carbon 145:694–700
32. Lin Y, Pan Y, Liu S, Sun K, Cheng Y, Liu M, Wang Z, Li X, Zhang J (2019) Construction of multi-dimensional core/shell Ni/NiCoP nano-heterojunction for efficient electrocatalytic water splitting. Appl Catal B 259:118039
33. Eissa DS, El-Hagar SS, Ashour EA, Allam NK (2019) Electrochemical nano-patterning of brass for stable and visible light-induced photoelectrochemical water splitting. Int J Hydrogen Energy 44(29):14588–14595
34. Xi X, Li J, Ma Z, Li X, Zhao L (2019) Enhanced water splitting performance of GaN nanowires fabricated using anode aluminum oxide templates. RSC Advances 9(26):14937–14943
35. Wang Y, Zhu Y, Afshar S, Woo MW, Tang J, Williams T, Kong B, Zhao D, Wang H, Selomulya C (2019) One-dimensional CoS 2–MoS 2 nano-flakes decorated MoO 2 sub-micro-wires for synergistically enhanced hydrogen evolution. Nanoscale 11(8):3500–3505
36. Sun B, Yang S, Guo Y, Xue Y, Tian J, Cui H, Song X (2019) Fabrication of molybdenum and tungsten oxide, sulfide, phosphide (MoxW1-xO2/MoxW1-xS2/MoxW1-xP) porous hollow nano-octahedrons from metal-organic frameworks templates as efficient hydrogen evolution reaction electrocatalysts. J Colloid Interface Sci 547:339–349
37. Yang Y, Yao H, Yu Z, Islam SM, He H, Yuan M, Yue Y, Xu K, Hao W, Sun G (2019) Hierarchical nanoassembly of MoS2/Co9S8/Ni3S2/Ni as a highly efficient electrocatalyst for overall water splitting in a wide pH range. J Am Chem Soc 141(26):10417–10430

Chapter 3
Fossil Hydrocarbon Decarbonization and Nanotechnology

3.1 Fossil Hydrocarbon Decarbonization

Sustainable hydrogen can be produced from a variety of substances other than water such as fossil fuels (natural gas, oil derivatives, coal, and other hydrocarbons) [1]. Fossil fuels can serve as the source of sustainable hydrogen because of their price, availability, and convenience in transportation and storage. Carbon and hydrogen are the major constituents of the hydrocarbons, and hence serve as the promising raw material for the generation of hydrogen. Hydrogen can be generated by a variety of different methods from hydrocarbons and their derivatives. These processes are commonly termed as fossil hydrocarbon decarbonization. Some of the commercially employed fossil hydrocarbon decarbonization methods are discussed below.

Steam reforming of natural gas is the most effective and the commonly employed method for the generation of hydrogen. This process involves the catalytic conversion of methane (which is a chief constituent of the hydrocarbon feed) and water (steam) into the oxides of carbon and hydrogen. This process is generally operated at very high pressure (35 atm) and temperature (850–950 °C). The chemical reaction for the process can be given as.

$$CH_4 + 2H_2O \rightarrow 4H_2 + CO_2 + \text{energy}.$$

Another important method for hydrogen generation from fossil hydrocarbons is partial oxidation. In partial oxidation, oxygen, fuel, and sometimes steam are mixed in fixed ratios which result in the conversion of fuel into a mixture of oxides of carbon and hydrogen. The partial oxidation of hydrocarbons is used with several alterations depending on the type of composition of the substrate fossil hydrocarbons. The overall process is exothermic, and the reaction can proceed with or without catalysts. In the presence of catalyst, this reaction proceeds at lower temperatures (600–900 °C), whereas in the absence of the catalyst the reaction requires very high temperature (1100–1500 °C) to proceed along with the production of high molecular weight coal

S. Farrukh et al., *Nanotechnology and the Generation of Sustainable Hydrogen*, Green Energy and Technology, https://doi.org/10.1007/978-3-030-60402-8_3

and oils. The reaction can be given by the following reactions:

$$CH_4 + 1/2O_2 \rightarrow CO + 2H_2 + \text{energy}.$$
$$CH_4 + O_2 \rightarrow CO_2 + 2H_2 + \text{energy}.$$

Steam iron process is also used for the production of hydrogen from a large range of hydrocarbon. Even coal can be used as the raw material for hydrogen generation by using this process. Reduction–oxidation regenerative system is used for the generation of ultrapure hydrogen with different catalysts. Modified form of steam iron process also has fuel cell applications. This method requires high temperature and involves several steps.

Photocatalytic activation of hydrocarbons also results in the generation of H_2. The specific photocatalysts activate the hydrocarbon molecule by absorbing the near-UV photons of the solar spectrum. The photocatalytic activation involves photoinduced charge transfer in photoactive catalysts, which lead to the formation of active electron-deficient entities which extract the proton from hydrocarbon molecule leading to the generation of hydrogen molecule and other by-products.

Thermal decomposition is another process in which hydrocarbons are decomposed into hydrogen and carbon. Pyrolysis temperature of ~ 400 °C is required for thermal decomposition, and this much high temperature is achieved from fuel–air flame while the air being cut off. At these conditions, the hydrocarbons get pyrolyzed into hydrogen and carbon-black particles, as thermal decomposition is a cyclic process and was formally employed for the production of hydrogen. The major problem in production of hydrogen from the fossils fuels is the emission of CO_2. The concentration of CO_2 would increase gradually in the earth atmosphere due to increased energy demands. Nevertheless, of the procedure employed for the generation of hydrogen from hydrocarbons all the carbon (present in the hydrocarbon) will be converted into oxides of the carbon and mostly CO_2. This is the major challenge toward using hydrocarbons for the production of sustainable hydrogen from hydrocarbons [2].

The decarbonization of fossil hydrocarbons cannot be discussed without the production of CO_2. This chapter is focused on the use of nanotechnology in the hydrogen generation from hydrocarbons and CO_2 reduction, which is produced during the process of decarbonization of fossil fuels. A simplified scheme of hydrogen generation from fossil hydrocarbons is shown in Fig. 3.1: Simplified scheme of hydrogen generation from fossil hydrocarbons.

3.2 Steam Reforming

Several studies have reported the use of nanotechnology in steam reforming of the hydrocarbons and their derivatives. In a study, $Ni/Cu/Al_2O_3$ nanocatalysts were employed for the steam reforming of natural gas for the production of hydrogen.

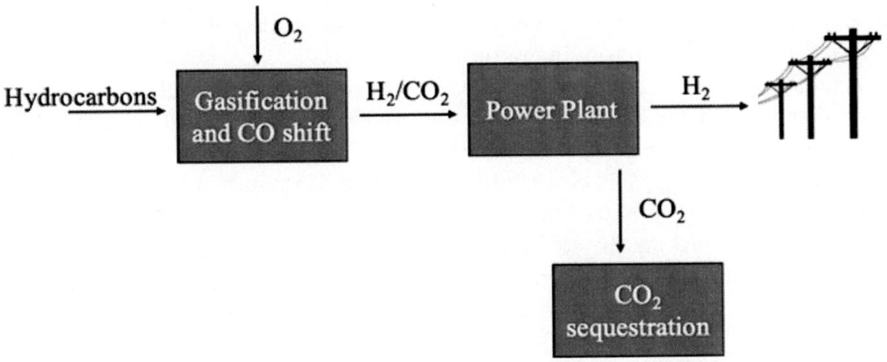

Fig. 3.1 Simplified scheme of hydrogen generation from fossil hydrocarbons

The study has shown that with the increase in temperature along with the introduction of additional steam increase the production of the hydrogen gas from the shale gas over Al_2O_3 sustained Ni/Cu nanocatalysts. In this study, the nanocatalyst was prepared by impregnation and co-precipitation method. Al $(NO_3)_3 \cdot 9H_2O$ was used for obtaining Al_2O_3 where Cu and Ni were obtained from Cu $(NO_3)_3 \cdot 3H_2O$ and Ni $(NO_3)_2 \cdot 6H_2O$, respectively. In impregnation method, the Al_2O_3 was impregnated and calcined prior to the addition of Ni and Cu, whereas in co-precipitation method all the reactants were first dispersed in water and later precipitated with K_2CO_3. The catalytic activity of the catalysts prepared by impregnation method was higher as compared to the other [3].

Ethanol is a derivative of ethane and can be used for the production of sustainable hydrogen by catalytic steam reforming. The major issue faced in this process is catalytic deactivation. *Chunfei* et al. have reported the steam reforming of ethanol by employing nano-Ni/SiO_2 catalyst. Sol –gel and impregnation methods were used for the preparation of the catalyst. For sol–gel method, Ni $(NO_3)_2.6H_2O$ and citric acid were dissolved in ethanol, and more ethanol was added afterward along with the DI water which is followed by the dropwise addition of tetraethyl silicate. In case of impregnation method, Ni $(NO_3)_2 \cdot 6H_2O$ was dissolved in deionized (DI) water and mixed with silica, and later it was dried and calcined in the air for 3 h. All the prepared catalysts (both methods) were reduced in hydrogen atmosphere at 600–700 °C. The catalysts were characterized with scanning electron microscope (SEM) coupled with energy-dispersive X-ray spectrometer (EDXS), Brunauer–Emmett–Teller (BET), and thermal gravimetric analysis (TGA). A fine dispersion of Ni and large BET surface areas of $> 700\ m^2\ g^{-1}$ were achieved for sol−gel catalysts, whereas for impregnation catalyst small BET surface area of $1\ m^2\ g^{-1}$ was obtained. The catalyst prepared by sol–gel method generated double the amount of hydrogen as compared to the impregnation catalysts. This can be explained by the fact that the catalyst obtained by sol–gel method showed uniform dispersion of Ni, whereas irregular dispersion of the metal was observed in the catalysts prepared by impregnation method. The sol–gel catalyst was also found to be stable than the other when used repeatedly in the steam reforming of the ethanol [4].

Another study has reported use of Al_2O_3/NiO nanocatalyst in quartz for the steam reforming of methane. The catalyst was prepared by employing sol–gel method. It was found that the size of NiO crystallite in NiO–SiO_2/Al_2O_3 catalyst varies with the degree of Ni loading and calcination temperature. The study determined that the nanocatalyst with 10% Ni is finest for steam reforming of methane. The catalyst was prepared by making the solution of tetraethyl orthosilicate (TEOS) in ethanol and nickel nitrate hexahydrate (amount depending upon the required Ni loading) in aqueous ethanol. Afterward, both solutions were slowly mixed together with constant stirring. The solution was allowed to stir at 25 °C for ~ 5–6 h. The product thus obtained was aged for 7 days. After aging, the catalyst was dried and calcined at specific temperature. For the introduction of Al_2O_3, NiO–SiO_2 and alumina were mixed together, and bentonite was added as the powder. SEM, transmission electron microscope (TEM), X-ray diffraction (XRD), BET, temperature-programmed reduction (TPR), and thermal conductivity detector (TCD) were used for the characterization of the product. The catalyst is found to be stable up to 500–700 °C. At optimized conditions, 95.7% conversion of methane to hydrogen was achieved [5].

Keshavarz et al. reported Ni/$MgAl_2O_4$ nanocatalysts for the steam pre-reforming of natural gas. In this investigation, the catalysts were prepared by the deposition precipitation process, which involves the surfactants. Different parameters which effect the activity of catalyst and the dispersion of the Ni like aging time, nature of surfactant, precipitation temperature pH, and of solution were also studied. The best conditions for the preparation of catalysts were found to be the 5 h aging time, sodium stearate as the surfactant, 30 °C temperature, and pH 10. At 500–550 °C, 100% ethane and propane conversions were achieved. For the preparation of the catalyst, firstly, $MgAl_2O_4$ was prepared by co-precipitation method by using polyvinylpyrrolidone (PVP) and capping agent. Later on, 21% Ni loading was achieved via deposition precipitation. For this purpose, surfactant and Ni $(NO_3)_2 \cdot 6H_2O$ were stirred together in DI water followed by the addition of $MgAl_2O_4$ and NaOH. Later on, the product was dried and calcined at high temperature. Different characterization techniques were employed for the analysis of the catalysts like H_2-TPR, XRD, BET, N_2 adsorption/desorption, TPR, inductively coupled plasma optical emission spectrometer (ICP), temperature-programmed oxidation (TPO), and temperature-programmed desorption (TPD) [6].

3.3 Partial Oxidation

Nanotechnology has also found its applications in partial oxidation of the hydrocarbons and their derivatives. Several studies have reported the partial oxidation of different hydrocarbons for generation of sustainable hydrogen by employing certain nanostructures. A novel catalyst consisting of three-dimensional honeycomb-like SiO_2 enclosed by ZrO_2 layer was reported to assist the partial oxidation of methane.

This ZrO_2/SiO_2 was synthesized by sol–gel coating process. The prepared material has average diameter of ~ 10 nm. The silica structure was prepared with pluronic, trimethylbenzene, and tetraethoxysilane. The prepared honeycomb-like assembly of the silica was modified with zirconium n-propoxide, and glacial acetic acid, in order to add the ZrO_2 encasing. Afterward, dispersed Ni nanoparticles were incorporated on ZrO_2/SiO_2 assembly by impregnation method by using Ni $(NO_3)_2 \cdot 6H_2O$. The prepared catalysts were analyzed with N_2 adsorption–desorption, TEM, SEM, H_2-TPR/TPD, TGA, and XRD. The fabricated nanocatalyst depicted better coking and anti-sintering ability in comparison to the traditional counterparts. The rate of methane conversion by employing honeycomb ZrO_2/SiO_2-Ni was found to be 90–92%. This superior performance of these catalysts is associated with the better metal dispersion in three-dimensional honeycomb SiO_2, thermal stability of the Ni nanoparticles, and the multiple interfaces between Ni, ZrO_2, and honeycomb support [7].

Marin et al. have reported the partial oxidation of Jet-A fuels to hydrogen and carbon monoxide via nanoparticle of MoO_2. These nanoparticles were prepared by the reduction of MoO_3 in 1:3 by volume solution of ethylene glycol and distilled water. The prepared nanoparticle MoO_2 was analyzed with XRD, SEM, BET, and X-ray photoelectron spectroscopy (XPS). MoO_2 nanoparticles depicted ~99% conversion of the fuel (at 850 °C, 1 atm) and ~60% yield of hydrogen. In the similar study, the MoO_2 catalyst was compared with the reference Ni catalyst at similar conditions which was deactivated because of coking in less than 4 h of the process where MoO_2 nanoparticles showed a 10-h coking resistance [8].

Ni nanoparticles are known for their ability to partially oxidize methane to generate hydrogen. In a study conducted in 2019, the stability of the Ni nanocatalysts was enhanced by using the reactive oxygen species (ROS) in certain metal oxides. The Ni nanoparticles were immobilized inside the different mesoporous metal oxides such as Yb_2O_3, La_2O_3, CeO_2, and ZrO_2. This was done via evaporation-induced self-assembly method by utilizing honeycomb-like silica as substrate. The honeycomb-like silica structure was synthesized with pluronic, trimethylbenzene, and tetraethoxysilane, and the mesoporous oxides were formed with propoxide of the particular metal oxide, glacial acetic acid, and hydrochloric acid. Later on, the Ni catalyst was immobilized on them by wet impregnation method via Ni $(NO_3)_2 \cdot 6H_2O$. The highly activated nanoparticles were characterized with TEM, SEM, and H_2-TPR. The metal oxide-supported Ni nanocatalysts showed superior catalytic stability, and this stability can be attributed to the strong interaction between Ni and metal oxides, high dispersion of the active metal, and the confinement effect. By using these enhanced catalysts, 90–92% methane conversion was achieved. Abundant hydrogen production was reported by the use of this improved Ni nanocatalyst via partial oxidation of methane [9].

The catalytic activity of the Pt nanoparticles was improved by confining them on pores of silicalite-1 zeolite. *Lichao* et al. synthesized these improved Pt nanoparticles for the partial oxidation of methane to syngas (a mixture of fuel gases with hydrogen as the major component). The improvement in the ability of the nanoparticles was achieved by crystal transformation of Pt/silicalite-1 zeolite. Parent silicalite-1 zeolite was synthesized by stirring together the tetrapropylammonium hydroxide, tetraethyl orthosilicate, and water at 80 °C. The product thus obtained was later autoclaved. Metal nanoparticles supported on the silicalite-1 zeolite surface were then prepared by incipient-wetness impregnation. This was achieved by making the solution of $H_2PtCl_6 \cdot 6H_2O$ in DI water. This solution was then added to the vacuum dried prepared silicalite-1 zeolite. The hollow Pt/silicalite-1 zeolite was prepared by adding Tetra-n-propylammonium hydroxide in product followed by ultrasonication and autoclave placement at 170 °C. The prepared nanoparticles were characterized with XRD, H_2-TPR, SEM, N_2/Ar adsorption–desorption, CO adsorption, TGA, TEM, and ICP. The improved nanoparticles showed remarkable conversion of CH_4 with significant reduction in the diffusion path of the chemical reaction [10].

In a study, production of hydrogen by the partial oxidation of methanol has been reported by using the Au nanoparticles maintained on the metal oxides. H_2 and CO_2 were obtained as the major product of the reaction, whereas CO as minor product and H_2O as the by-product in sufficient quantity. Metal oxides like Fe_2O_3–MO_x were used where M can be Al, Zn, or Zr. The catalysts were synthesized in two steps. The first step involves the formation of composite oxide supports by impregnation of the Al_2O_3, ZrO_2, or ZnO support on the aqueous solution of Fe $(NO_3)_3 \cdot 9H_2O$. Later on, Au was dispersed on the fabricated support materials by deposition–precipitation process. The catalytic activity was shown by Au/Fe_2O_3-Al_2O_3 (conversion: 100%, H_2 selectivity: 48%), whereas the minimum activity was shown by Au/Fe_2O_3- ZrO_2. Al_2O_3 was found to be responsible in preventing the Au nanoparticles against sintering through calcination as verified from TEM analysis. With the increase in calcination temperature the concentration of the metallic Au in the catalyst increases. The pretreatment of the catalyst was also found to have effect on its activity, for instance, un-calcined Au/Fe_2O_3-Al_2O_3 depicted greater selectivity for H_2, whereas the calcined Au/Fe_2O_3-Al_2O_3 has shown poor performance in this regard. The synthesized catalysts were analyzed with XPS, XRD, TGA, and TEM [11].

One of the major problems faced during partial oxidation of hydrocarbons for hydrogen generation is carbon deposition and catalyst deactivation because of sintering. A bimetallic nanocatalyst Ni-Mo immobilised on ceria-zerconia was prepared by *Bkour* et al. and used for the partial oxidation of isooctane. These catalysts are found to cope with the problems of carbon deposition and deactivation. The catalyst was prepared by co-impregnation process. They were employed as proficient catalyst for partially oxidizing isooctane. The prepared material was used to form a bi-layer anode for micro-reforming on a traditional solid oxide fuel cell. The bimetallic catalysts were characterized with TCD, TEM, XRD, TPR, energy-dispersive spectroscopy (EDS), and scanning transmission electron microscopy (STEM). The introduction of Mo in the catalysts increases the conversion of the substrate to syngas, along with increase in stability and dispersion of Ni catalysts. Carbon tolerance of the Ni was also found to be increased with increased concentration of Mo. These nanocatalysts depicted 75% yield of H_2 and 100% conversion in 24 h by partial oxidation of isooctane [12].

3.4 Photocatalytic Activation

Photocatalytic activation of different hydrocarbons and their derivatives for generation of H_2 involves the extensive use of nanotechnology.

VO_2 is recognized as semiconductor with 0.7 eV bandgap, and this ability makes it a promising candidate for the photocatalysis. *Wang* et al. reported a body-centered-cubic, nanostructured VO_2, with remarkable photocatalytic activity for the generation of hydrogen from ethanol and water mixture. This ability of the VO_2 nanostructures can be attributed to their very high optical bandgap (~2.7 eV). These nanostructures were prepared by thermal oxidation process; the nanostructures of VO_2 were deposited on Si substrate. The Si substrate was selected because it has low substrate temperature for producing extensive arrays of aligned nanorods and nanostructures of VO_2 (also for other metal oxides) on its surface. The prepared nanostructures were studied with SEM, TEM, XRD, electron energy loss spectrometry (EELS), Rutherford backscattering spectrometry (RBS), and selected area (electron) diffraction (SAD). The VO_2 nanostructures depicted the quantum yield of ~38.7% when fabricated and utilized as nanorods. The production rate of hydrogen can be improved by adjusting the UV-incident angle on VO_2 nanorods (aligned in the form of film). By correctly adjusting the angle of incident UV light with power density of ~27 m W/cm^2, a high hydrogen production rate of 800 $mmol/m^2h^{-1}$ is achieved [13].

Pt/TiO_2 (B) nanofibers are noted for their photocatalytic activity in the production of hydrogen from hydrocarbons. In a study, they are used for the generation of H_2 from neat ethanol. For the synthesis of Pt/TiO_2 (B) nanofibers, firstly, nanofibers of $H_2Ti_3O_7$ were synthesized via hydrothermal process in NaOH at 130 °C followed by acidic wash and subsequent drying at 110 °C. Afterward, $H_2Ti_3O_7$ nanofibers were calcined at 300 °C which resulted in the development of TiO_2 (B) nanofibers. Later the Pt was impregnated to these fibers. It was observed that with the increase in calcined temperature by 100 °C, the crystallinity of the TiO_2 (B) nanofibers improved whereas the dimension and morphology of the fibers remained same. The nanofibers were characterized with field-emission scanning electron microscope (FE-SEM), TEM, and Raman spectroscopy. The FE-SEM shows that many of the $H_2Ti_3O_7$ and TiO_2 (B) nanofibers are tightly held together in the form of fiber bundle. The study showed that the catalyst which was calcined at 400 °C has the H_2 production rate similar to the standard P-25 TiO_2 whereas the catalysts which were calcined at above or below 400 °C had shown comparatively poor results [14].

TiO_2 nanofilms decorated with Pt are also involved in the photocatalytic generation of H_2 from methanol gas. TiO_2 films were developed with tetrabutyl orthotitanate in ethanol followed by the addition of chlorohydric acid. The films were fabricated by spin coating method on the surface of quartz and calcined at 400 °C.

Pt was incorporated on the films by infusing them in solution of H_2PtCl_6 for 6 h. The excessive H_2PtCl_6 was removed by treating the films with formaldehyde. The films were characterized with XRD and SEM. The photocatalytic degradation of methanol gas was carried out in continuous-flow tank with Pt/TiO_2 films. In this process, maximum H_2 production rate of 4.675 mmolh^{-1} was achieved [15].

Hydrogen can be generated from triethanolamine via the photocatalytic activation using eosin Y-sensitized Pt-loaded nanotubes $Na_2Ti_2O_4(OH)_2$. The photocatalytic activity of eosin Y-sensitized Pt-loaded is reported by *Li* et al. under visible light irradiation with wavelength $\geq$ 420 nm. These nanotubes were synthesized by the hydrothermal synthesis process whereas the eosin Y-sensitized NTS was synthesized by the impregnation process. Pt nanoparticles were by impregnated in situ with H_2PtCl_6 by photodeposition process. The prepared catalysts were analyzed by TEM, solid-state diffusion reflectance UV–vis spectrometer (UV–vis/DRS), and XRD. The catalyst exhibited 14.97% quantum efficiency, 100 h stability (in 10 consecutive rounds) [16].

In an investigation, a photocatalytic bi-crystalline mixture of two nanostructures was used for the production of hydrogen from neat ethanol. For the preparation of first set of nanostructures, nanotubes of sodium titanate were transformed into nanotubes of monoclinic TiO_2 (B) by washing with HCl and then drying at 300 °C. When the calcination temperature of prepared nanotubes was increased to 400 °C these nanotubes start to change into anatase nanoparticles resulting in the formation of bi-crystalline combination consisting of nanotubes of TiO_2 (B) and nanoparticles of anatase. The prepared nanostructures were characterized with XRD and FE-SEM. Scherrer equation was used to determine the principal particle size of the anatase nanoparticles, which were assessed to be ~10 nm. When the bi-crystalline structures were loaded with 1% Pt, their photocatalytic ability was improved drastically. They generated 25% more H_2 by photocatalytic dehydrogenation in UV light as compared to the reference P-25 TiO_2. The promising efficiency of the bimetallic catalysts can be attributed to the fact that the photocatalyst which consists of two phases provides better separation of photogenerated electrons and holes, leading to the better photocatalytic activity [17].

3.5 CO_2 Sequestration

Fossil decarbonization for the production of sustainable hydrogen always produces oxides of the carbon relative to the amount of carbon present in the hydrocarbon. The CO_2 is produced as the major oxides. This is one of the biggest disadvantages of the fossil hydrocarbon decarbonization. Several solutions have been proposed for the mitigation of the problem. Nanotechnology is also involved in many of these solutions.

Li et al. have reported the simultaneous generation of H_2 and the carbon nanotubes from the fossil decarbonization of ethanol over Fe/Al_2O_3 catalyst. For the decarbonization of ethanol, fresh Fe/Al_2O_3 catalyst was placed in quartz tube (at constant temperature zone) inside fixed reactor. The temperature of catalysts was then increased up to 500 °C and followed by the flow of Ar and H_2 for 1 h at rate of 40 $mlmin^{-1}$. H_2 flow is turned off after reduction, whereas Ar is allowed to flow again. After attaining the constant temperature, ethanol is introduced in the gasification room at controlled rate. Here decomposition of ethanol takes place, resulting in the liberation of hydrogen gas and carbon deposition on the surface of the catalyst. Different characterization techniques such as Raman spectroscopy, TEM, and SEM proved that the deposited materials on the surface of the catalysts are carbon nanotubes. Carbon nanotubes thus formed not only reduce the CO_2 emission but also have several advantages of their own in multiple fields [18].

The CO_2 emitted during decarbonization of hydrocarbons can also be converted into single-walled carbon nanotubes over Fe/MgO catalyst. In an investigation, a derivative of Feitknecht-compound precursor is used as catalyst. The catalyst was synthesized with co-precipitation process. The Feitknecht-compound precursor is known for the development of strong metal–support association which allows the development of fine quality single-walled carbon nanotubes on its surface during decarbonization of ethanol for hydrogen production. TEM was used for the surface characterization of the prepared carbon nanotubes [19].

3.6 Conclusion

The hydrocarbons can be used for the production of hydrogen. The decarbonization of hydrocarbons is achieved by using different processes such as partial oxidation, photocatalytic activation, steam reforming, etc. The evolution of hydrogen from hydrocarbons is accompanied with the production of carbon dioxide. Nanomaterials are aiding both the liberation of hydrogen and mitigation of carbon dioxide, which is produce as the result of decarbonization of the hydrocarbons. The current study proposed the use of comparatively less consumed fractions from the obtained crude oil refining, instead of already useful hydrocarbon such as natural gas. Extensive research is needed in this area. A very little research is done in the area of carbon dioxide mitigation, obtained as by-product during decarbonization. Along with the conversion of carbon in carbon nanotubes, other mitigation processes must also be used for handling of the gas. As only the conversion of generated carbon dioxide is not enough for large-scale decarbonization of the hydrocarbons.

References

1. Mueller-Langer F, Tzimas E, Kaltschmitt M, Peteves S (2007) Techno-economic assessment of hydrogen production processes for the hydrogen economy for the short and medium term. Int J Hydrogen Energy 32(16):3797–3810. https://doi.org/10.1016/j.ijhydene.2007.05.027

2. Muradov N (2002) Hydrogen from Fossil Fuels without Co^2 Emissions. In: Grégoire Padró CE, Lau F (eds) Advances in hydrogen energy. Springer US, Boston, MA, pp 1–16. doi:https://doi.org/10.1007/0-306-46922-7_1
3. Jabarullah NH, Othman R (2019) Steam reforming of shale gas over Al2O3 supported Ni-Cu nano-catalysts. Pet Sci Technol 37(4):386–389
4. Wu C, Williams PT (2010) A novel nano-Ni/SiO2 catalyst for hydrogen production from steam reforming of ethanol. Environ Sci Technol 44(15):5993–5998. https://doi.org/10.1021/es100912w
5. Bej B, Pradhan NC, Neogi S (2013) Production of hydrogen by steam reforming of methane over alumina supported nano-NiO/SiO2 catalyst. Catal Today 207:28–35. https://doi.org/10.1016/j.cattod.2012.04.011
6. Keshavarz AR, Soleimani M (2016) Optimization of nano-sized Ni/MgAl2O4 catalyst synthesis by the surfactant-assisted deposition precipitation method for steam pre-reforming of natural gas. RSC Advances 6(66):61536–61543. https://doi.org/10.1039/C6RA07166J
7. Guo S, Wang J, Ding C, Duan Q, Ma Q, Zhang K, Liu P (2018) Confining Ni nanoparticles in honeycomb-like silica for coking and sintering resistant partial oxidation of methane. Int J Hydrogen Energy 43(13):6603–6613. https://doi.org/10.1016/j.ijhydene.2018.02.035
8. Marin-Flores O, Turba T, Ellefson C, Scudiero L, Breit J, Norton MG, Ha S (2010) Nanoparticle Molybdenum Dioxide: A New Alternative Catalytic Material for Hydrogen Production via Partial Oxidation of Jet-A Fuels. J Nanoelectron Optoelectron 5(2):110–114. https://doi.org/10.1166/jno.2010.1074
9. Ding C, Wang J, Guo S, Ma Z, Li Y, Ma L, Zhang K (2019) Abundant hydrogen production over well dispersed nickel nanoparticles confined in mesoporous metal oxides in partial oxidation of methane. Int J Hydrogen Energy 44(57):30171–30184. https://doi.org/10.1016/j.ijhydene.2019.09.202
10. Ma L, Ding C, Wang J, Li Y, Xue Y, Guo J, Zhang K, Liu P, Gao X (2019) Highly dispersed Pt nanoparticles confined within hierarchical pores of silicalite-1 zeolite via crystal transformation of supported Pt/S-1 catalyst for partial oxidation of methane to syngas. Int J Hydrogen Energy 44(39):21847–21857. https://doi.org/10.1016/j.ijhydene.2019.06.051
11. Roselin LS, Liao L-M, Chang F-W (2017) Gold Nanoparticles Supported on Fe2O3–MO x (M = Al, Zr, Zn) Composite oxides for partial oxidation of methanol. J Nanosci Nanotechnol 17(4):2796–2803. https://doi.org/10.1166/jnn.2017.12725
12. Bkour Q, Zhao K, Scudiero L, Han DJ, Yoon CW, Marin-Flores OG, Norton MG, Ha S (2017) Synthesis and performance of ceria-zirconia supported Ni-Mo nanoparticles for partial oxidation of isooctane. Appl Catal B 212:97–105. https://doi.org/10.1016/j.apcatb.2017.04.055
13. Wang Y, Zhang Z, Zhu Y, Li Z, Vajtai R, Ci L, Ajayan PM (2008) Nanostructured VO2 Photocatalysts for Hydrogen Production. ACS Nano 2(7):1492–1496. https://doi.org/10.1021/nn800223s
14. Lin C-H, Chao J-H, Liu C-H, Chang J-C, Wang F-C (2008) Effect of calcination temperature on the structure of a Pt/TiO2 (B) nanofiber and its photocatalytic activity in generating H2. Langmuir 24(17):9907–9915. https://doi.org/10.1021/la800572g
15. Cui W, Feng L, Xu C, Lü S, Qiu F (2004) Hydrogen production by photocatalytic decomposition of methanol gas on Pt/TiO2 nano-film. Catal Commun 5(9):533–536. https://doi.org/10.1016/j.catcom.2004.06.011
16. Li Q, Lu G (2007) Visible-light driven photocatalytic hydrogen generation on Eosin Y-sensitized Pt-loaded nanotube Na2Ti2O4(OH)2. J Mol Catal A: Chem 266(1):75–79. https://doi.org/10.1016/j.molcata.2006.10.047
17. Kuo H-L, Kuo C-Y, Liu C-H, Chao J-H, Lin C-H (2007) A highly active bi-crystalline photocatalyst consisting of TiO2 (B) nanotube and anatase particle for producing H2 gas from neat ethanol. Catal Lett 113(1):7–12. https://doi.org/10.1007/s10562-006-9009-1
18. Li W, Wang H, Ren Z, Wang G, Bai J (2008) Co-production of hydrogen and multi-wall carbon nanotubes from ethanol decomposition over Fe/Al2O3 catalysts. Appl Catal B 84(3):433–439. https://doi.org/10.1016/j.apcatb.2008.04.026

19. Jin Y, Wang G, Li Y (2012) Catalytic growth of high quality single-walled carbon nanotubes over a Fe/MgO catalyst derived from a precursor containing Feitknecht compound. Appl Catal a 445–446:121–127. https://doi.org/10.1016/j.apcata.2012.08.010

Chapter 4
Hydrogen Sulfide Decomposition and Nanotechnology

4.1 Hydrogen Sulfide Decomposition

Hydrogen sulfide is yet another important source of sustainable hydrogen. H_2S is categorized as a highly toxic pollutant of both natural and anthropic origins. Naturally, Black Sea can be regarded as the one major reservoir of H_2S because of its internal structure. The upper layer is aerobic and runs down to the length of 150 m. The lower layer consists of the anoxic sea where the concentration of the compound increases and with the increase in depth, it is ultimately reached to the constant concentration of ~9.5 mg/L. Sulfur-reducing bacteria are responsible for this high concentration of H_2S [1]. Coal seams can be considered as the second major natural source of the H_2S. The presence of the compound in the coal mines is also responsible for the large number of deaths of the mine workers. The presence of large quantity of H_2S is either due to the thermochemical reactions or because of the bacterial sulfate reduction [2]. Different anthropic activities are the source of the H_2S generation, mainly petrochemical and oil extraction processes. The crude oil is enriched with large quantity of sulfur-containing compounds including disulphides, thioethers, polybenzothiophenes, etc. and their derivatives. Hydrodesulfurization of these compounds in the process of purifying crude oil results in the formation of H_2S [3, 4]. H_2S is produced as a dangerous chemical effluent from various industries, hence presenting a serious threat of poisoning to the industrial workers [5]. H_2S is also produced in the biomass gasification plants as sulfur is present in the combined form in the feedstocks [6]. It can also be produced from the wastewater treatment because of anaerobic oxidation of the dissolved sulfur compounds in the feed [7].

H_2S can be used as the promising source for the generation of sustainable hydrogen. The decomposition of H_2S serves two important purposes. Firstly, it eliminates one of the most toxic pollutants, and secondly it generated the sustainable H_2. There are several methods for the decomposition of H_2S for the liberation of H_2. Some of them are briefly discussed here.

S. Farrukh et al., *Nanotechnology and the Generation of Sustainable Hydrogen*, Green Energy and Technology, https://doi.org/10.1007/978-3-030-60402-8_4

Photocatalytic dissociation of H_2S liberates the hydrogen, in the presence of some suitable photocatalyst and light. The photodissociation of H_2S normally occurred in liquid phase. The particular photocatalysts stimulate the substrate molecule by attracting the near-UV light of the spectrum. This stimulation involves the photo-induced transfer of charge in photocatalysts, which generated dynamic electron-deficient species which extract the proton from H_2S molecule leading to the production of H_2 molecule and other products. The chemical reactions of H_2S photodissociation in aqueous medium can be given as.

$$H_2S + OH \leftrightarrow SH^- + H_2O.$$

$$H_2S + 2OH^- \leftrightarrow S^{2-} + 2H_2O.$$

$$2SH^- \rightarrow S2^{2-} + 2H^+$$

$$2H^+ + 2e^- \rightarrow H_2.$$

Thermolysis is another technique for the decomposition of hydrogen sulfide. This technique is commonly employed due to its simplicity and easy operational conditions. In this technique, the decomposition of H_2S is carried out at high temperatures. The thermal decomposition of H_2S occurs at varying temperatures depending upon the applied pressure and initial concentration of the substrate [8]. The digramatic explaination of pyrolysis of H_2S is shown in Fig. 4.1.

Thermocatalytic methods involve the use of both temperature and the catalyst for the liberation of the H_2 from H_2S. Claus process is normally used for thermal catalytic cracking of H_2S. The choice of the catalysts for this process depends upon the advantages of short residence time and material resistance. At low optimal pressures, the commonly employed catalysts for the process are sulfides of transition metals like Mo, W, V, Fe, Co, Ni, Cu, and Zn [9]. Bimetallic catalysts like sulfides of Ru–Mo, Cd

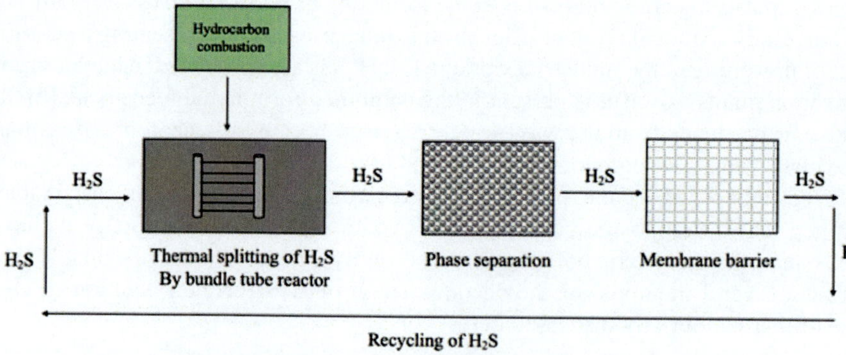

Fig. 4.1 Diagrammatic explanation of pyrolysis of H_2S

chalcogens, activated carbon, and perovskites of transition and rare earth metals are also used for the purpose. Fluidized catalysts have shown better catalytic activities as compared to the bed catalysts due to the enhanced cleaning effect on the surface of catalyst and improved intraparticle mass transfer [10–12].

H_2S molecule can also be dissociated into sulfur and sustainable hydrogen via electrochemistry technique. In this process, the hydrogen is liberated at cathode (reduction reaction) where sulfur is obtained at anode (oxidation reaction), and the deposition of the sulfur at anode also results in the severe passivation of anode. Therefore, heat is used in the electrochemical assembly for the transfer of sulfur into sulfur steam [13]. The chemical reactions for the process can be given as.

$$H_2S + 2e^- \rightarrow H_2 + S^{2-} \quad \text{(reduction half reaction).}$$

$$S^{2-} \rightarrow S + 2e^- \quad \text{(oxidation reaction)}$$

$$H_2S^- \rightarrow H_2 + S \quad \text{(overall reaction)}$$

Multistep thermochemical processes are also available for the decomposition of H_2S to generate H_2. These processes involve two-step mechanism depending upon H_2S adsorption (via a single metal or group of metals) and methods that are taken from water splitting reactions. The well-known multistep thermochemical processes for decomposition of H_2S are sulfurization of metals, inorganic sulfides or oxides, thermochemical sulfur–iodine cycles, and revamping of CO or COS cycles [14–16].

Several H_2S decomposition processes involve the use of nanotechnology. The common nanomaterial involved in the dissociation of H_2S molecule involves the transition metal-based nanomaterials such as nanoparticles of different Cd chalcogens. Some carbon- based nanomaterials such as graphene oxide (GO) powder also involve in the decomposition of the compound. The current chapter will explain the frequently used nanomaterials in H_2S decomposition for the generation of sustainable hydrogen.

4.2 Cadmium-Based Nanomaterials and Hydrogen Sulfide Decomposition

Cd salts with different materials, particularly sulfur, are used as the catalyst for the conversion of H_2S to H_2. However, they are susceptible to photo-corrosion. Different studies have suggested different solutions to the problem. In a study, the nano-CdS powder was synthesized in hetero-matrix solid-state polymer-inorganic reaction. The nano-CdS powder eliminated the problem of photo-corrosion along with the improved photocatalytic ability due to the presence of charge transfer. These thermally stable nanocrystallites of CdS restrained on polyphenylene sulfide (PPS) matrix were prepared by using cadmium iodide and PPS. The mixture of the two compounds was heated at 285 °C (melting point of PPS) in a Pt crucible for 48 h. The product thus obtained was analyzed with X-ray diffractions (XRD), transmission electron microscopy (TEM), and field-emission scanning electron microscope (FESEM). The characterization of the product showed that the CdS nanocrystallites of 6 to 28 nm in size were restrained on PPS matrix. 19.7% quantum yield of H_2 production was achieved with the prepared nanomaterials [17].

Yao et al. have reported a facile method for the production of Pd–Cr_2O_3/CdS nanocomposite to be used as a photocatalyst for the generation of H_2 from H_2S. For the preparation of the nanocomposites, the aqueous solution of H_2PdCl_4 and Cr $(NO_3)_3$. nH_2O was allowed to react with $NaBH_4$. The accumulation of the nanoparticles during H_2PdCl_4 reduction was prevented by the addition of polyvinylpyrolidone (PVP) in the solution with thorough mixing. Afterward, CdS was added to the solution with vigorous stirring resulting in the formation of Pd–Cr_2O_3/CdS nanocomposite. A quantum yield of H_2 increased from 37.3% to 55.6% at 420 nm by optimizing chromium oxide loading of CdS in nanocomposite catalyst. The prepared catalysts depicted better photocatalytic ability as compared to the pristine Pd. This can be associated to the presence of interface for charge transfer. No characterization analysis of the prepared composites was reported in the study [18].

Nanostructure of spinel semiconductors of varying morphologies has huge significant applications as photocatalyst. A group of investigators synthesized a cubic spinel chemically stable nanostructure of $CdIn_2S_4$ via simple hydrothermal process as visible light active photocatalyst for the generation of hydrogen by decomposing H_2S. For the preparation of these nanostructures, mixture of Cd $(NO_3)_2 \cdot 4H_2O$, $In(NO_3)_3 \cdot 3H_2O$, and excessive thiourea was placed in stainless-steel autoclave lined with teflon along with double distilled water at 140 °C for 60 h. The product was obtained in form of yellow precipitates. The obtained product was given ethanol wash and was dried at 70 °C. The marigold morphology was shown by the samples in aqueous medium whereas nanotubes with 25 nm diameter were formed in organic solvent (methanol) under similar conditions. Different characterization techniques like XRD, TEM, and FESEM were used for the analysis of the prepared nanostructures. The nanocatalyst exhibited excellent photocatalytic activity in both mediums. The $CdIn_2S_4$ nanostructures showed give hydrogen quantum yields of 17.1% with nanotube morphology whereas 16.8% with marigold-like morphology in visible light. The promising ability of the catalyst for the generation of hydrogen is directly associated to the high crystallinity of the prepared catalyst [19].

A quantum dot-glass nanosystem of $CdS_{0.5}Se_0$ and CdSe was prepared and used for the hydrogen production from hydrogen sulfide by *Sanjay* et al. The synthesized nonensemble was grown on germanate glass matrix via facile melt quench technique. The glass used for the preparation of the nanosystem consists of GeO_2, Na_2O K_2O, ZnO, TiO_2, and $B_2O_3.Al_2O_3$. The CdSe and $CdS_{0.5}Se_05$ were used as dopants on the glass surface. The nanosystem was prepared by making the adequately homogenized mixture of host glass powder and the dopant CdS0.5Se0.5/CdSe. Electrical furnace was used for melting of the mixture at 900–950 °C for 2 h. Later on, the product was quenched and annealed at 425 °C in an adjustable furnace. Afterward, the product is allowed to grow on the pieces of blank glass at 425 °C–475 °C for 8 h. The increase in temperature caused the increase in the size of quantum dots of CdS0.5Se0.5/CdSe as the growth is dependent on temperature because of nucleation and growth mechanisms. XRD and Raman spectroscopic analysis of the sample revealed that the $CdS_{0.5}Se_{0.5}$ and CdSe are monodispersed in quantum dot-glass nanosystem and have hexagonal structure. CdSe quantum dots have size in the range of 2–12 nm. TEM and UV–Vis spectroscopy were used to determine the quantum confinement of $CdS_{0.5}Se_{0.5}$ and CdSe. The control of $CdS_{0.5}Se_{0.5}$ quantum dot size allowed the adjustment of bandgap from 3.6 eV to 1.8 eV. This improved the absorption of visible light, which resulted in the enhanced catalytic activity. 21 and 26% apparent quantum yields of hydrogen liberation were achieved by using CdSe and $CdS_{0.5}Se_{0.5}$ quantum dot–glass nanosystems, respectively, for the decomposition of H_2S [20].

4.3 Titanium-Based Nanomaterials and Hydrogen Sulfide Decomposition

Titanium-based photocatalyst is frequently used in the decomposition of H_2S for the generation of H_2. In a study, a nanocomposite was fabricated with bulk CdS and nanocrystalline TiO_2 as the photocatalyst for the generation of hydrogen from aqueous solution of H_2S under visible light of the solar spectrum. For the preparation of the CdS, aqueous solution of Na_2S was mixed with $Cd(NO_3)_2$ and further dissolved in isopropyl alcohol. A precipitated product was obtained which was calcined at 800 °C for 1 h. The calcination was performed in helium atmosphere so as to enhance the crystallinity of CdS. For the synthesis of CdS/TiO_2 bulk-nanocomposite photocatalyst, the prepared CdS was added to isopropyl alcohol and titanium isopropoxide, water was added slowly afterward, and the mixture was stirred. The composite formed was calcined at 800 °C for 2 h in the flow of air for enhancing the TiO_2 crystallinity. The prepared composite was characterized with XRD, TEM, and UV–Vis diffuse reflectance spectrometer. The study suggested that the use of the prepared catalysts is helpful in hydrodesulfurization plants and Claus plant gas steams for the simultaneous production of H_2 and removal of H_2S [21, 22].

In an investigation, a titanium dioxide film-coated reactor was made via heat treatment and sol–gel process as the potential candidate for the photocatalytic degradation of H_2S. The prepared compound depicted excellent H_2S degradation capability for the generation of H_2. Tetrabutyl titanate was used as the precursor for the synthesis the TiO_2. Solution of tetrabutyl titanate was made in absolute ethanol followed by the addition of a mixture containing distilled water, glacial acetic, and absolute ethanol. This resulted in the formation of semi-transparent sol, which was allowed to age for 7 days. A quartz pipe (cleaned with sulfuric acid, distilled water, and ultrasonication) was used as the base for the deposition of TiO_2 film. The prepared sol was discharged into the quartz pipe (the pipe was wrapped with plastic from outside), and this resulted in the deposition of the film onto the inner surface of the pipe. Six layers were applied in the similar manner, and the pipe was dried for 5 min before the application of the next layer. Afterward, the TiO_2-coated quartz pipe was calcined at 500 °C for 2 h. This quartz pipe coated with TiO_2 film was used as the photochemical reactor. The prepared films were analyzed in XRD, X-ray photoelectron spectroscopy (XPS), and SEM. From the characterization, it was found that the TiO_2 film with mean crystalline size of 24 nm was formed [23].

Nilima et al. have developed nanostructures of N-doped TiO_2 in marigold-like morphology with solvothermal process for H_2 generation from H_2S. The N-doped TiO_2 was prepared making the solution of titanium tetra-isopropoxide in methanol followed by the addition of hydrazine hydrate. This resulted in the formation of white precipitate indicating the formation of complex between Ti^{4+} and hydrazine hydrate. Later on, solution of guanidine carbonate in acetic acid was added in the mixture, which dissolved the formed complex (white precipitates). The whole mixture was then placed in the solution in sealed autoclave with Teflon lining at 150° C for 9–16 h. The product thus obtained was washed and dried at 80 ° C. The structural examination of the prepared nanomaterials has shown that the Ti is present in anatase form in the synthesized nanostructures. The N-doping in the nanostructures is confirmed by photoluminescence and photoemission spectroscopy. The microscopic analysis has shown that the thickness of petal in N–TiO_2 marigold-like nanostructures is around ~2–3 nm. These nanostructures have been proven to be effective photocatalyst for the dissociation of H_2S for the generation of hydrogen under sunlight [24].

4.4 Zinc-Based Nanomaterials and Hydrogen Sulfide Decomposition

Zinc-based photocatalysts are also involved in the decomposition of the H_2S for evolution of H_2. Nanostructured $ZnIn_2S_4$ has been found to be very active in the photocatalytic decomposition of H_2S for the production of sustainable hydrogen. In an investigation, hydrothermal process was employed for the fabrication of $ZnIn_2S_4$ nanoparticles. The said nanostructures were developed by placing the mixture of Zn $(NO_3)_2.6H_2O$, $In(NO_3)_3.5H_2O$, and excessive thiourea and double distilled water in stainless-steel autoclave lined with Teflon, along with double distilled water at 150 °C for 30 h. Yellow precipitates were obtained as product. The obtained product was given ethanol wash and was dried at 70 °C. The marigold morphology of the prepared nanostructures was studied with FESEM, and it was found that the nanomaterials have marigold-like morphology with the size of flower in the range of 3–5 μ m and the thickness of the petal was ~ 3–5 nm. 5287 μ mol/h evolution of hydrogen was achieved by utilizing these nanostructures via the decomposition of H_2S under visible light radiations. The chemical reaction for the evolution of hydrogen with the said catalyst can be given as [25].

$$H_2S + OH^- \leftrightarrow SH^- + H_2O.$$

$$ZnIn_2S_4 \rightarrow h^+ + e^- \quad (h^+ = \text{valence band hole; } e^- = \text{conduction band electron}).$$

$$2SH^- + 2h^+ \rightarrow S_2^{2-} + 2H^+ \quad (\text{oxidation reaction}).$$

$$2H^+ + 2h^+ \rightarrow S_2^{2-} + 2H^+ \quad (\text{oxidation reaction}).$$

The core–shell nanomaterials with the core of TiO_2 and shell of CdS–ZnS nanoparticles were prepared for the generation of H_2 via decomposition of H_2S. TiO_2 nanotubes were developed by heating the mixture of titania powder, water, and sodium hydroxide at 110 °C for 20 h. The nanotubes were attained by washing the prepared product with 6% of nitric acid, isopropanol, and water for the removal of sodium ions. The nanotubes were then converted into nanorods by placing the aqueous solution of nanotubes in autoclave at 175 °C for 48 h. The CdS-ZnS/TiO_2 core–shell nanoparticles were prepared via co-precipitation process. In a typical method, thiourea was added to the solution of cadmium acetate and zinc acetate, and afterward Na_2S and synthesized TiO_2 nanorods were gradually added and the whole mixture was stirred for ~15 h. This resulted in the formation of crystallized nanoparticles which were recovered by ultracentrifugation. Later on, the prepared nanoparticles were washed with deionized (DI) water and isopropyl alcohol and dried in oven at 110 °C. The core–shell nanoparticles were characterized with XRD, TEM, SEM, FESEM, Brunauer–Emmett–Teller (BET), and XPS. The prepared nanomaterials were also analyzed for the determination of specific surface area, molecular vibrations, particle size, bandgap energy, and binding energy. When CdS–ZnS/TiO_2 core–shell nanoparticles were used as catalyst for production of hydrogen, from aqueous solution of sulfide and sulfite ions, an evolution rate of 29 mL/h was observed. Under optimized conditions, maximum 30% conversion was obtained [26].

An investigation has reported green synthesis of p-type N-doped ZnO nanostructures with the ability to decompose H_2S into sustainable hydrogen under sunlight. The catalyst was prepared via wet chemical method. In the typical method, solution of $ZnCl_2$ in absolute ethanol was prepared and urea was added to it with continuous stirring. Later on, the solution was placed in oven at 150 °C and a white hygroscopic product was formed. The product was preserved in a desiccator. Afterward, the product was annealed at high temperatures ranging from 500 °C to 800 °C for 3 h. This resulted in the formation of non-hygroscopic brown products. After annealing the product was washed with hot distilled water for the removal of excessive chlorine. The structural analysis of the p-type N-doped ZnO showed that the lattice of the said material is hexagonal wurtzite. The N-doping of the nanomaterials was confirmed by optical studies which indicted a severe shift in bandgap from 3.19 eV to 2.3 eV in the visible region of the spectrum. The evidence of N-doping of the ZnO catalyst was also supported by Raman scattering and XPS analysis. The p-type N-doped ZnO were characterized morphologically with FESEM, TEM. The particle size of the nanostructures was in the range of ~4–5 nm. The nanostructure showed ~3957 μ mol/h rate of hydrogen evolution [27]. The chemical reaction for the evolution of hydrogen with p-type N-doped ZnO from H_2S can be given as.

$$H_2S + OH^- \leftrightarrow SH^- + H_2O.$$

$$N - ZnO \rightarrow h^+ + e^- \quad (h^+ = \text{valence band hole};\ e^- = \text{conduction band electron}).$$

$$2SH^- + 2h^+ \rightarrow 2S^{2-} + 2H^+ \quad (\text{oxidation reaction}).$$

$$2H^+ + 2e^- \rightarrow H_2 \quad (\text{reduction reaction}).$$

4.5 Other Nanomaterials and Hydrogen Sulfide Decomposition

In an investigation, hierarchical nanostructure-like dandelion flowers and nanorods of Bi_2S_3 were developed by solvothermal process. The Bi_2S_3 nanomaterials were developed making the solution of Bi $(NO)_3 \cdot 5H_2O$ and thiourea in a mixture of water and ethylene glycol and were stirred for 1 h. The whole mixture was then placed in stainless-steel autoclave lined with Teflon at 150 °C. After the completion of the reaction, the system was allowed to cool at room temperature. Black precipitates were obtained in the form of product which were washed with water and ethanol at 80 °C. FESEM, XRD, and TEM analysis were used for the characterization of the synthesized nanomaterials. The prepared nanomaterials were utilized for the photocatalytic decomposition of H_2S for H_2 generation under visible light. The prepared Bi_2S_3 hierarchical nanostructure depicted remarkable hydrogen generation of 8.88 mmol/g/h under daylight (between 11.30 am and 2.30 pm on bright sunny day) whereas the nanorods generated 7.08 mmol/g/h of hydrogen under similar conditions from the decomposition of H_2S. This can be associated to the fact that the hierarchical nanostructure has multiple surface defects which improve the separation of the charge carriers [28].

In another study, $CuGaO_2$ and its indium-doped analogue $CuGa_{1-0.065}In_{0.065}O_2$ delafossite oxides co-catalysts nanostructures were fabricated and utilized for the generation of hydrogen from H_2S. The nanocatalysts were prepared via solid-state process. XRD analysis of the materials depicted that they have hexagonal rhombohedral structure. The morphological analysis with FESEM of the nanocatalysts indicated that $CuGaO_2$ has deformed plate-like particles while $CuGa_{1-0.065}In_{0.065}O_2$ have systematic hexagonal rod-like arrangements. The prepared catalysts were able to decompose hydrogen sulfide in aqueous KOH solution in visible light radiations. At the incident radiation of $\lambda = 550$ nm, quantum yield of 13.6% for hydrogen

Table 4.1 Nanomaterial and generation of the H_2 from H_2S

Material	Nanotechnology	Function	References
CdS in HY zeolite pores	Nanoparticles	Photocatalytic decomposition of H_2S	[30]
Oxynitride $Nb_2Zr_6O_{17-x}N_x$	Nanoparticles	Photocatalytic decomposition of H_2S	[31]
CdS sensitized $CdWO_4$	Nanoparticles grown on nanorods	Photocatalytic decomposition of H_2S	[32]
CdS	Nanosized and capsule	Photocatalytic decomposition of H_2S	[33]

generation was achieved. The study used following formula for the determination of quantum yield [29]:

$$\text{Quantum yield} = \frac{2x\text{Number of Hydrogen molecules evolved}}{\text{Number of incident photon}} x100$$

Some other nanostructures for the generation of the H_2 from H_2S have been reported which are listed in Table 4.1.

4.6 Conclusion

Hydrogen sulfide, which is a major environmental pollutant can be uses as the potent source of hydrogen. In this way, not only the harmful pollutant is eliminated but also clean energy is generated. Transition metal-based nanomaterials are actively involved in the liberation of hydrogen from hydrogen sulfide. The major problem is the capturing of the hydrogen sulfide gas from various natural and anthropogenic sources. The modern researchers should focus on developing the systems which assist the efficient capturing of H_2S from various sources, so that considerable amount of hydrogen can be obtained from H_2S.

References

1. Naman SA, Veziroğlu A (2013) The low cost hydrogen production from hydrogen sulfide in Black Sea. In: Black sea energy resource development and hydrogen energy problems. Springer, Berlin, pp 93–108
2. Hoşgörmez H, Yalçın MN, Soylu C, Bahtiyar İ (2014) Origin of the hydrocarbon gases carbon dioxide and hydrogen sulfide in Dodan Field (SE-Turkey). Mar Pet Geol 57:433–444
3. Jarullah AT, Mujtaba IM, Wood AS (2011) Kinetic parameter estimation and simulation of trickle-bed reactor for hydrodesulfurization of crude oil. Chem Eng Sci 66(5):859–871
4. Zuckerman JJ, Hagen AP (1987) Inorganic Reactions and Methods, Vol. 2. The Formation of the Bond to Hydrogen (Part 2). VCH Publishers, Weinheim, Germany,

5. Zhao Y, Biggs TD, Xian M (2014) Hydrogen sulfide (H 2 S) releasing agents: chemistry and biological applications. Chem Commun 50(80):11788–11805
6. Emami-Taba L, Irfan MF, Daud WMAW, Chakrabarti MH (2013) Fuel blending effects on the co-gasification of coal and biomass–a review. Biomass Bioenergy 57:249–263
7. Mahmood Q, Zheng P, Cai J, Hayat Y, Hassan MJ, Wu D-l, Hu B-l (2007) Sources of sulfide in waste streams and current biotechnologies for its removal. J Zhejiang Univ-Sci A 8(7):1126–1140
8. Hawboldt KA, Monnery WD, Svrcek WY (2000) New experimental data and kinetic rate expression for H2S pyrolysis and re-association. Chem Eng Sci 55(5):957–966
9. Zaman J, Chakma A (1995) Production of hydrogen and sulfur from hydrogen sulfide. Fuel Process Technol 41(2):159–198
10. Cao D, Adesina AA (1999) Fluidised bed reactor studies of H2S decomposition over supported bimetallic Ru catalysts. Catal Today 49(1–3):23–31
11. Guldal NO, Figen HE, Baykara SZ (2015) New catalysts for hydrogen production from H2S: preliminary results. Int J Hydrogen Energy 40(24):7452–7458
12. Maloka IE, Aliwi SM, Naman SA (2006) Thermal decomposition of hydrogen sulphide by cadmium chalcogens. Pet Sci Technol 24(1):103–112
13. Weaver D, Winnick J (1987) Electrochemical Removal of H 2 S from Hot Gas Streams: Nickel/Nickel-Sulfide Cathode Performance. J Electrochem Soc 134(10):2451
14. Gibson AG, Wachs IE Hydrogen generation from petroleum refinery off-gas. In: 2006 Amer chemical soc 1155 16TH ST, NW, WASHINGTON, DC 20036 USA,
15. Kiuchi H, Nakamura T, Funaki K, Tanaka T (1982) Recovery of hydrogen from hydrogen sulfide with metals or metal sulfides. Int J Hydrogen Energy 7(6):477–482
16. Leybros J, Carles P, Borgard J-M (2009) Countercurrent reactor design and flowsheet for iodine-sulfur thermochemical water splitting process. Int J Hydrogen Energy 34(22):9060–9075
17. Kanade KG, Baeg J-O, Mulik UP, Amalnerkar DP, Kale BB (2006) Nano-CdS by polymer-inorganic solid-state reaction: Visible light pristine photocatalyst for hydrogen generation. Mater Res Bull 41(12):2219–2225. https://doi.org/10.1016/j.materresbull.2006.04.031
18. Yao W, Huang C, Muradov N, T-Raissi A, (2011) A novel Pd–Cr2O3/CdS photocatalyst for solar hydrogen production using a regenerable sacrificial donor. Int J Hydrogen Energy 36(8):4710–4715. https://doi.org/10.1016/j.ijhydene.2010.12.124
19. Kale BB, Baeg JO, Lee SM, Chang H, Moon SJ, Lee CW (2006) CdIn2S4 nanotubes and "Marigold" nanostructures: a visible-light photocatalyst. Adv Func Mater 16(10):1349–1354
20. Apte SK, Garaje SN, Naik SD, Waichal RP, Baeg J-O, Kale BB (2014) Quantum confinement controlled solar hydrogen production from hydrogen sulfide using a highly stable CdS 0.5 Se 0.5/CdSe quantum dot–glass nanosystem. Nanoscale 6(2):908–915
21. Jang JS, Li W, Oh SH, Lee JS (2006) Fabrication of CdS/TiO2 nano-bulk composite photocatalysts for hydrogen production from aqueous H2S solution under visible light. Chem Phys Lett 425(4):278–282. https://doi.org/10.1016/j.cplett.2006.05.031
22. Preethi V, Kanmani S (2013) Photocatalytic hydrogen production. Mater Sci Semicond Process 16(3):561–575. https://doi.org/10.1016/j.mssp.2013.02.001
23. Yu Y, Zhang T, Zheng L, Yu J (2013) Photocatalytic degradation of hydrogen sulfide using TiO2 film under microwave electrodeless discharge lamp irradiation. Chem Eng J 225:9–15. https://doi.org/10.1016/j.cej.2013.03.032
24. Chaudhari NS, Warule SS, Dhanmane SA, Kulkarni MV, Valant M, Kale BB (2013) Nanostructured N-doped TiO 2 marigold flowers for an efficient solar hydrogen production from H 2 S. Nanoscale 5(19):9383–9390
25. Chaudhari NS, Bhirud AP, Sonawane RS, Nikam LK, Warule SS, Rane VH, Kale BB (2011) Ecofriendly hydrogen production from abundant hydrogen sulfide using solar light-driven hierarchical nanostructured ZnIn2S4 photocatalyst. Green Chem 13(9):2500–2506
26. Ruban P, Sellappa K (2016) Concurrent hydrogen production and hydrogen sulfide decomposition by solar photocatalysis. Clean—Soil, Air, Water 44(8):1023–1035. https://doi.org/10.1002/clen.201400563

27. Bhirud AP, Sathaye SD, Waichal RP, Nikam LK, Kale BB (2012) An eco-friendly, highly stable and efficient nanostructured p-type N-doped ZnO photocatalyst for environmentally benign solar hydrogen production. Green Chem 14(10):2790–2798
28. Kawade UV, Panmand RP, Sethi YA, Kulkarni MV, Apte SK, Naik SD, Kale BB (2014) Environmentally benign enhanced hydrogen production via lethal H 2 S under natural sunlight using hierarchical nanostructured bismuth sulfide. RSC Adv 4(90):49295–49302
29. Gurunathan K, Baeg J-O, Lee SM, Subramanian E, Moon S-J, Kong K-J (2008) Visible light assisted highly efficient hydrogen production from H2S decomposition by CuGaO2 and CuGa1−xInxO2 delafossite oxides bearing nanostructured co-catalysts. Catal Commun 9(3):395–402. https://doi.org/10.1016/j.catcom.2007.07.021
30. Bai X-f, Cao Y, Wu W (2011) Photocatalytic decomposition of H2S to produce H2 over CdS nanoparticles formed in HY-zeolite pore. Renewable Energy 36(10):2589–2592. https://doi.org/10.1016/j.renene.2010.04.037
31. Kanade KG, Baeg J-O, Kale BB, Mi Lee S, Moon S-J, Kong K-j (2007) Rose-red color oxynitride Nb2Zr6O17-xNx: A visible light photocatalyst to hydrogen production. Int J Hydrogen Energy 32(18):4678–4684. https://doi.org/10.1016/j.ijhydene.2007.07.040
32. Sethi YA, Panmand RP, Kadam SR, Kulkarni AK, Apte SK, Naik SD, Munirathnam N, Kulkarni MV, Kale BB (2017) Nanostructured CdS sensitized CdWO4 nanorods for hydrogen generation from hydrogen sulfide and dye degradation under sunlight. J Colloid Interface Sci 487:504–512. https://doi.org/10.1016/j.jcis.2016.10.063
33. Takahashi H, Yokoyama S, Baba Y, Hayashi T, Tohji K Effective hydrogen generation from the hydrogen sulfide solution by using stratified type photocatalyst. In: 2008 American Institute of Physics, pp 17–21

Chapter 5
Biomass Decomposition and Nanotechnology

5.1 Biomass Decomposition

Biomass can be used as the potential alternative for the generation of sustainable hydrogen. It is not considered the best candidate to be used as the feedstock for the generation of hydrogen because of very low hydrogen content (~6% against 25% of methane). The content of H_2 is smaller in biomass due to large (40%) oxygen content. Despite the low hydrogen contents biomass is still used as the important source for the generation of sustainable hydrogen as it is renewable and absorbs atmospheric CO_2, which makes it a better option than the fossil fuels [1] because the fossils fuels are finite and depleting. Fossil fuels also produce considerable amount of CO_2 during the generation of H_2 resulting in global warming and other environmental issues [2]. Biomass is obtained through variety of sources such as crop residues, waster of agricultural industries, and plant residues like wood and the crops [3]. There are several methods for the generation of H_2 from biomass. A diagrammatic classification of various hydrogen generation methods is given in Fig. 5.1. These methods can be broadly classified into two categories, i.e., thermochemical processes and biochemical processes [4]. Biochemical processes are appropriate for the biomass enriched with starch or sugar but do not work well with the lignocellulosic-based biomass. Contrary to this, thermochemical processes are suited for the broad range of biomass feedstocks [5, 6]. Some of the important methods for the generation of H_2 from biomass are discussed here.

Fast pyrolysis of biomass is a thermochemical process in which H_2 is produced from biomass. This kind of pyrolysis results in the formation of small amount of gaseous hydrogen and pyrolytic oil is also referred as bio-oil, which is later on converted to hydrogen by steam reforming. Fast pyrolysis is usually followed by steam reforming for the recovery of the H_2 from the pyrolytic oil. The oil is separated into two fractions depending upon the miscibility in water, namely, the water-soluble fraction and the water-insoluble fraction. The water-soluble fraction is used for the liberation of hydrogen whereas the other part cannot be employed for the purpose

S. Farrukh et al., *Nanotechnology and the Generation of Sustainable Hydrogen*, Green Energy and Technology, https://doi.org/10.1007/978-3-030-60402-8_5

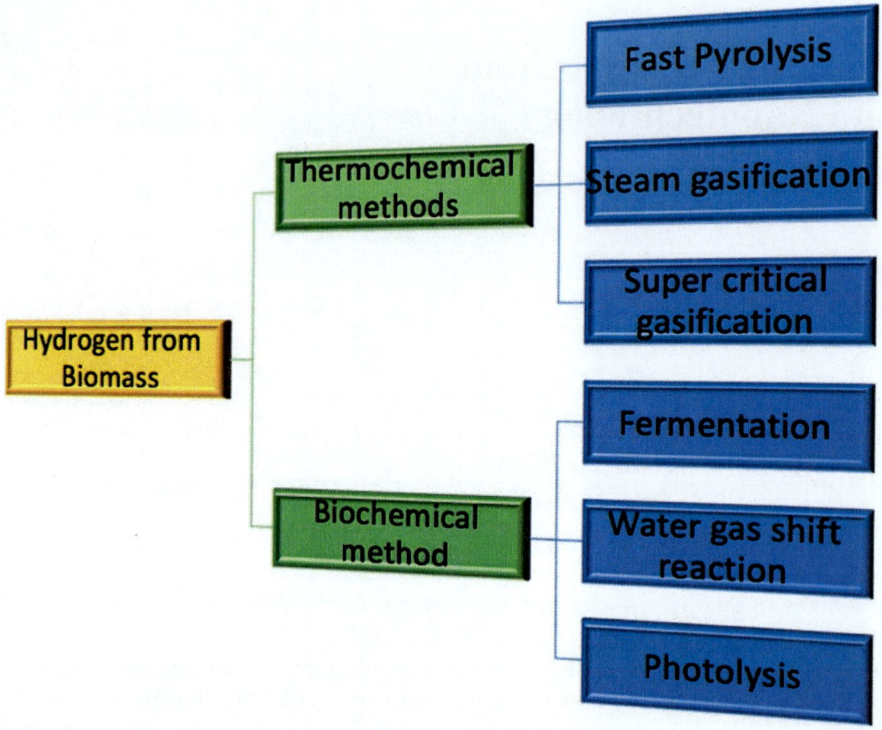

Fig. 5.1 Various processes for generation of hydrogen from biomass

but can be used in the preparation of the adhesives. There are some disadvantages associated with this process such as low efficiency and production of by-product like char and tar. Char contaminates the gaseous products where tar is responsible for the clogging of the equipment pipes as it is an oily emulsion [7].

Gasification is another thermochemical process which generates H_2 from biomass. This process utilizes different materials as the gasifying medium for the production of hydrogen gas such as steam, air, and oxygen. The gasification of the biomass is always preceded by pyrolysis as the rate of the later is greater than gasification process. Pyrolysis results in the formation of char (residue) when it reacts with gasifying agent. Steam gasification is considered as a remarkable technique for the generation of H_2-rich fuel gas. The composition of the average fuel gas generated by steam gasification is stated to be 40% H_2, 8% CH_4, 25% CO, 25% CO_2, and 2% N_2. On the weight basis of biomass, the theoretical yield of H_2 is calculated to be 17% maximum. Gasification can produce syngas with 50–55% of H_2 [8–11]. Gasification of the biomass involves a series of reactions during the generation of H_2 from biomass. The important reactions are given below [12–16]:

Major Pyrolysis Reaction

$$\text{Biomass} \rightarrow \text{char} + H_2 + CO + CO_2 + CH_4 + \text{hydrocarbon} + \text{tar}$$

Oxidation Reactions

$$C + 1/2O_2 \rightarrow CO + 1/2O_2 \rightarrow CO_2$$

$$H_2 + 1/2O_2 \rightarrow H_2O$$

Carbon Gasification (Boudouard Reaction)

$$C + CO_2 \rightarrow 2CO$$

Methane Formation

$$C + 2H_2 \rightarrow CH_4$$

Water Gas Reactions

$$C + H_2O \rightarrow CO + H_2$$

$$C + 2H_2O \rightarrow CO_2 + 2H_2$$

Steam Reforming Reactions

$$CH_4 + H_2O \rightarrow CO_2 + 3H_2$$

$$CH_4 + H_2O \rightarrow CO_2 + 3H_2$$

Carbon Monoxide Shift Reaction

$$CO + H_2O \rightarrow H_2$$

Cracking Reactions

$$C_XH_Y + nH_2O \rightarrow n\,CO_2 + (n + 1/2)H_2$$

$$\text{Tar} + H_2O \rightarrow CO_2 + H_2 + CO + \text{low molecular weight hydrocarbons}$$

Supercritical water gasification (SWG) is another thermochemical process for hydrogen production. The gas and liquid phases of water become completely miscible when treated at supercritical conditions (>22.1 MPa and > 374 °C). In these conditions, the oxygen in the water behaves as an oxidant. When the supercritical water is allowed to react with the biomass (a carbonaceous material) it will oxidize the carbon of the biomass resulting in the generation of CO. After further oxidation, the CO results in the formation of CO_2. When carbon of the biomass is utilized, the H_2 and O_2 of both biomass and water were liberated. This method appears very promising but is very costly and is only suitable for biomass with high moisture content [17, 18].

Fermentation is common biological process for the generation of hydrogen from biomass. There are two types of fermentation for the evolution of hydrogen from biomass, i.e., the light fermentation and the dark fermentation. The light fermentation utilizes photoheterotrophic microorganisms whereas the dark fermentation makes use of the anaerobic microbes for the fermentation of the biomass. In the process of fermentation, the first step is hydrolysis (either enzymatic or acidic) of the biomass to produce the concentrated solution of sugars which is later on fermented to hydrogen, oxygen, and volatile fatty acids with anaerobic organisms. The organic acids present in the concentrate are additionally fermented by the photoheterotrophic bacteria such as *Rhodobacter sp.* and liberate H_2 and CO_2. When light and dark fermentation are used collectively greater amount of H_2 is generated from the carbohydrates [19]. The chemical reaction for the conversion of carbohydrate to H_2 by fermentation can be given as.

$$C_6H_{12}O_6 + 2H_2O_2 \rightarrow 2CH_3COOH + 2CO_2 + 4H_2$$

$$C_6H_{12}O_6 \rightarrow CH_3(CH_2)_2 \cdot COOH + 2CO_2 + 2H_2$$

Biomass is also converted to H_2 via photosynthesis process. Different phototropic microbes like cyanobacteria, green bacteria, and purple bacteria and numerous algae are utilized for the generation of hydrogen from biomass in the presence of sunlight. Green algae and other microbes absorb the sunlight and produce electrons, and the photosystem sends these ferredoxin electrons.

Biological water gas shift reaction is yet another biochemical process in which the particular microbes like *Rubrivivax gelatinosus* are used for the conversion of the biomass to sustainable hydrogen. These microbes are capable of carrying out water gas shift reactions with biomass at optimized temperature and pressure. The general reaction of the process can be given as [20].

$$\text{Biomass} \rightarrow H_2O + CO \rightarrow CO_2 + H$$

Both the thermochemical and the biological processes of biomass conversion for the generation of sustainable H_2 involve the extensive use of nanotechnology. Many studies have used the nanotechnology for the thermochemical conversion of the algal mass to the sustainable hydrogen [21]. Similarly, various studies of biomass fermentation have been reported which employ different nanomaterials of both inorganic and organic origins for the production of H_2. For instance, nanomaterials of different metal and metal oxides such as Cu, Au, Pd, Ag, and iron-iron-oxide. TiO_2 are involved in dark fermentation of the biomass for hydrogen generation [22]. The current chapter is based on the applications of nanomaterials that are used in H_2 generation from biomass.

5.2 Thermochemical Conversion of Biomass and Nanotechnology

Nanomaterials are commonly employed as catalysts in various thermochemical reactions for the generation of hydrogen from biomass, such as gasification, supercritical water gas formation, and steam reforming. In a study, trimetallic Ni-La-Fe/γ-Al_2O_3 nanomaterials were used for biomass gasification reaction. The nanomaterial was prepared for the removal of tar in biomass steam gasification for the production of improved quality gas. The said nanomaterials were prepared by deposition–precipitation (DP) method. For the preparation of the materials, γ-Al_2O_3 substrate was depressurized at few torr for 1 h. This was done for the removal of air from the pores of the substrate. On the other hand, solution of $Ni(NO_3)_2.6H_2O$, $La(NO_3)_3.6HO$, $Fe(NO_3)_3.9H_2O$, and urea was made in distilled water and stirred at room temperature until a homogenized mixture is formed. Afterward, the mixture was transferred on the γ-Al_2O_3 substrate which was sealed in the glass vessel and heated in the oil bath at 115 °C for 2.5 h. The product was obtained in the form of brown deposition on the surface of the substrate. After the completion of the reaction, the product was cooled at optimized conditions and washed with deionized (DI) water until the colorless product was obtained. The colorless nanomaterials were first dried at 90 °C for 6 h and then calcined at 550 °C for 1 h. Scanning electron microscope (SEM) and X-ray diffractometer (XRD) were used to determine radial cross section of the disk and morphology. The particle size was analyzed with transmission electron microscopy (TEM) and the metal content was determined by energy-dispersive spectroscopy (EDX). The catalyst activity of the prepared material was evaluated by conducting the gasification experiment with saw dust (as the biomass) with and without the nanocatalysts. When the catalysts were used, 99% removal of the tar and 10% volume increase in the generation of H_2 were observed [23].

In a study, biomass was used as the raw material for the production of biodiesel, and crude glycerol was obtained as the significant by-product. Afterward, hydrogen was produced from this crude glycerol in the presence Ni $(OH)_2$ decorated on TiO_2 nanotubes under sunlight. In the practice, the TiO_2 nanotubes were prepared first by hydrothermal method, and later on they were deposited by $Ni(OH)_2$ quantum dots via wet impregnation method. For this purpose, TiO_2 particles were dissolved in the solution of NaOH and placed in the autoclave lined with Teflon at 130 °C for 20 h. The white product thus obtained was washed with distilled water, dilute HCl, and ethyl alcohol. The washed product was calcined at 350 °C for 5 h. Afterward, Ni was loaded on the prepared nanotubes by using the aqueous solution of Ni $(NO_3)_2 \cdot 6H_2O$. The mixture was constantly heated while being stirred to remove the excessive solvent. At the end of the reaction, fine product was obtained which was dried at 80 °C and calcined at 350 °C for 3 h. The prepared product was analyzed with XRD and TEM. The characterization of the product showed that the TiO_2 nanotubes have length and diameter in the range of 100–300 nm and 4.9–9.8 nm, respectively. The quantum dots of $Ni(OH)_2$ decorated on nanotubes showed the average particle size of 8.4 nm. The XPS studies indicated that Ti and Ni are present in the oxidation state of Ti^{4+} and Ni^{2+} in the prepared nanomaterials. 4719 μ mol/h H_2 generation was achieved under solar irradiation from the biomass-based glycerol via the use of prepared nanomaterials [24].

Some studies have used different nanomaterials in supercritical water gasification of biomass for the generation of hydrogen. *Dennis* et al. have reported the use of Pt-carbon nanotubes in the generation of sustainable hydrogen from supercritical water reforming of acetic acid and ethylene glycol (biomass products). For the preparation of the nanomaterials, pre-made carbon nanotubes were used without any additional purification. $H_2PtCl_6.6H_2O$ was used for the loading of the Pt on the nanotubes with wet impregnation method. In the practice, the solution of $H_2PtCl_6.6H_2O$ was made in acetone and added to the carbon nanotubes. The excessive solvent is evaporated by placing the system in vacuum at 100 °C. Afterward, the nanocatalysts were subjected to 500 °C temperature for 15 h in reducing atmosphere of H_2 and N_2. The loading of the Pt on the nanotubes was determined by X-ray fluorescence spectrometer (XFS); surface area was determined by Brunauer–Emmett–Teller (BET); and the surface morphology of the pure nanotubes and the loaded nanotubes was determined by SEM, X-ray photoelectron spectrum (XPS), TEM, XRD, and Raman spectroscopy [25].

Ethylene glycol is a simple modeled compound of ethylene biomass-derived liquids. It can be used for the production of sustainable hydrogen via steam reforming. In an investigation, Ni-Pt bimetallic nanocatalysts were used for the steam reforming of ethylene glycol. The study reported 60% conversion of the hydrocarbon and 27% yield of the hydrogen at optimized conditions. The bimetallic catalysts were synthesized by the wet impregnation method. For this purpose, firstly the solutions were made from Ni $(NO_3)_2 \cdot 6H_2O$ and $H_2PtCl_6 \cdot 6H_2O$ and were later on mixed with γ-aluminum oxide and stirred in a rotary evaporator at 60 °C. The reaction was allowed to take place for 5 h. Afterward, the product was collected, dried (90 °C), and calcined (600 °C) for 6 h. The synthesized catalysts were analyzed with TEM, BET, XRD, H_2-chemisorption, and H_2-TPR. The introduction of the Pt in the lattice of the Ni increased the reducibility of the catalyst which leads to the improved metal dispersion, greater coking resistance, and better oxygen consuming activity. This leads to the better conversion of the hydrocarbon and improved yield [26].

5.3 Biochemical Conversion of Biomass and Nanotechnology

Several studies have reported the use of wide-ranging nanomaterials for the fermentation of biomass for the production of hydrogen. In a study, hematite nanoparticles were used for the fermentation of sucrose in order to generate hydrogen. The nanoparticles were prepared by using the solution of $FeCl_3$. The solution was placed in the oven at 100 °C for 72 h which resulted in the formation of red precipitates which were separated by centrifugation. The obtained product was washed with ethanol and water. Afterward, the synthesized nanoparticles were resuspended in aqueous media via ultrasonication. The prepared nanoparticles were characterized with TEM and XRD. The TEM analysis revealed that the hematite nanoparticles have the average diameter of ~ 55 nm. The XRD studies indicated that the nanoparticles are in hexagonal form. *Clostridium butyricum* was used for the fermentation, and the microbe was employed as inocula without additional pretreatments. The study reported that the maximum 3.57 mol of H_2 were obtained per mole of sucrose when 200 mg/L of the nanoparticles were used [27].

As mentioned earlier, water gas shift reaction is an important biochemical reaction for the generation of hydrogen from biomass. Different studies have employed varying nanomaterials in the water gas shift reaction for obtaining better yield and high-quality hydrogen. In a study, researchers have claimed the use of gold nanoparticles for the generation of hydrogen via water gas shift reaction. The study reported that the method of preparation and the pretreatment of the catalysts is very critical for the functioning of the catalyst in water gas shift reaction. In order to achieve the catalytic activity of the Au nanoparticles in the reaction, the close contact between oxide support and gold is crucial. The close contact is associated with the zerovalent atoms of Au which is necessary for the high catalytic activity. The abovementioned hypothesis is supported by the in situ extended X-ray absorption fine structure (EXAFS) measurements [28].

An investigation has determined the effects of zerovalent Fe and Ni nanoparticles against the divalent Fe^{2+} and Ni^{2+} ions on the proficiency of dark fermentation mesophilic reaction from glucose on the pretreated anaerobic sludge for the generation of hydrogen. The sludge was treated with heat-shock. For the fermentation experiment, glucose solution (5.5 pH) was taken in dark glass bottle and the supplemented metal particles (separately) were added in the varied concentrations. Afterward, the nutrient solution and inoculum were also added along with the particular amount of deionized water. The bottles were covered with butyl rubber cap. Before the determination of the results the bottles were also incubated. The zerovalent Fe nanoparticles depicted 8% increase in the hydrogen yield, and this is probably because of the fact that Fe^0 improved the activity of ferredoxin electron transfer and hydrogenase; however, contrary to the Fe nanoparticles, Ni nanoparticles showed insignificant improvement (~0.9%) in the generation of hydrogen [29].

In another study, nanoparticles of maghemite (magnetite + hematite) were used in the fermentation of starch wastewater for the generation of hydrogen. The magnetite nanoparticles were synthesized by making the collective solution of $FeCl_3 \cdot 6H_2O$ and $FeCl_2.4H_2O$ in distilled water. In a separate beaker, NaOH was heated at 80 °C and Ar was flown in it, and the formally prepared solution was added to it dropwise resulting in the formation of homogenized small particles. After the completion of the reaction, black precipitates were formed which were crystallize. Hematite nanoparticles were prepared by making the solution of $FeCl_3$ in distilled water. The solution was then placed in oven at 100 °C for 72 h. Afterward, the solution of magnetite nanoparticles and hematite nanoparticles was mixed together to obtain maghemite nanoparticles. The prepared maghemite nanoparticles were analyzed with SEM and particle dimensions were found to be in the range of 55 nm. When the nanoparticles were employed in the fermentation it was observed that the amount of hydrogen generated by employing maghemite nanoparticles along with the bacteria culture was greater (104.75 ± 12.39 mL-H_2/g) than the amount generated with the bacteria culture (66.22 ± 4.88 mL-H_2/g) alone. The improved production of hydrogen by using the maghemite nanoparticles is associated with their ability to enhance the bioactivity of hydrogen-generating microbes [30].

Zhidan et al. reported the use of carbon nanotubes for the generation of hydrogen form biomass via *Ruminococcaceae* fermentation. The study employed hydrogen producing reactors instead of microbial consortium with retained hydrogen product was used. Carbon nanotubes improved the retention of the microbes. Seed sludge with inactivated methanogenic bacteria was used as the biomass in the study. The biomass was treated with alkali for the adjustment of the pH. Pre-made carbon nanotubes with 12 nm diameter were used for the process. The study concluded that the carbon nanotubes noticeably intensified the biomass retention and reduced the startup time of the reaction. Carbon nanotubes also offered larger flocculation capacity to *Ruminococcaceae,* hence improving the yield of hydrogen [31].

Laurent et al. used metallic and metallic oxide nanoparticles restrained on the porous silica and added them on *Clostridium butyricum*-assisted biomass fermentation for hydrogen generation. For the preparation of nanoparticles, metallic salts of Ag, Cu, Pd, and Fe were used. The prepared nanoparticles were encapsulated inside the porous silica with co-gelation process. The process allows the doping of cations with inorganic matrix at the molecular scale in a single step. Congelation allows the simultaneous hydrolysis and condensation of two alkoxy silanes. All the metallic nanoparticles anchored in silica were calcined at a high temperature for the removal of organic moieties. After calcination as the final step, the products were reduced in hydrogen atmosphere. All the prepared nanoparticles showed greater production of H_2 as compared to the blank experiments and the nanoparticles without silica matrix. The Pd/SiO_2, Ag/SiO_2, Fe/SiO_2, and Cu/SiO_2 nanoparticles liberated ~95.8, ~93.8, ~119.4, and ~103.9 mL, respectively, of H_2 from the 5 g/L of glucose, where the referenced culture generated ~87 mL of H_2 [32].

There are many other studies which involve the use of nanotechnology in biomass conversion to hydrogen either by biochemical or thermochemical process. A brief account of the few is given in Table 5.1: Nanotechnology in biomass conversion to hydrogen.

Table 5.1 Nanotechnology in biomass conversion to hydrogen

Biomass	Method of H_2 generation	Nanotechnology	References
Artificial wastewater	Fermentation	Gold nanoparticles	[33]
Lignin	Gasification	Nanoparticle catalyst	[34]
Glucose	Fermentation	Ni nanoparticles	[35]
Cellulose	Catalytic hydrothermal conversion	SnO_2 and ZnO nanoparticles	[36]
Glucose	Fermentation	TiO_2 and magnetic hematite nanoparticle	[37]
	Gasification/pyrolysis	Nano-NiO/γ-Al_2O_3	[38]
Wastewater	Fermentation	Ag nanoparticles	[39]
Anaerobic sludge/sucrose	Fermentation	Mesoporous Fe_3O_4 nanoparticles	[40]

5.4 Conclusion

Hydrogen can also be obtained from biomass along with several other sources. There are different methods, such as thermochemical and biochemical, for the generation of hydrogen from biomass. Nanotechnology is used in almost all of these processes for the generation of hydrogen. Amount of hydrogen obtained from the biomass is very small as compared to other sources. The biomass with high moisture content must be investigated for better production of hydrogen. The advanced nanomaterials like graphene sheets and carbon nanotubes must also be investigated for the generation of hydrogen from a variety of biomasses.

References

1. Milne TA, Elam CC, Evans RJ (2002) Hydrogen from biomass: state of the art and research challenges. United States. https://doi.org/10.2172/792221
2. Ahmed I, Gupta AK (2009) Syngas yield during pyrolysis and steam gasification of paper. Appl Energy 86(9):1813–1821
3. White LP, Plaskett LG (1981) Biomass as fuel. Academic Press Ltd.
4. Kalinci Y, Hepbasli A, Dincer I (2009) Biomass-based hydrogen production: A review and analysis. Int J Hydrogen Energy 34(21):8799–8817. https://doi.org/10.1016/j.ijhydene.2009.08.078
5. Balat H, Kırtay E (2010) Hydrogen from biomass–present scenario and future prospects. Int J Hydrogen Energy 35(14):7416–7426
6. Ozbay N, Pütün AE, Pütün E (2006) Bio-oil production from rapid pyrolysis of cottonseed cake: Product yields and compositions. Int J Energy Res 30(7):501–510
7. Evans R, Boyd L, Elam C, Czernik S, French R, Feik C, Philips S, Chaornet E, Parent Y (2003) Hydrogen from biomass-catalytic reforming of pyrolysis vapors.
8. Jarungthammachote S, Dutta A (2012) Experimental investigation of a multi-stage air-steam gasification process for hydrogen enriched gas production. Int J Energy Res 36(3):335–345

9. Karmakar MK, Datta AB (2011) Generation of hydrogen rich gas through fluidized bed gasification of biomass. Biores Technol 102(2):1907–1913
10. Kaushal P, Tyagi R (2012) Steam assisted biomass gasification—an overview. Canadian J Chem Eng 90(4):1043–1058
11. Saxena RC, Seal D, Kumar S, Goyal HB (2008) Thermo-chemical routes for hydrogen rich gas from biomass: a review. Renew Sustain Energy Rev 12(7):1909–1927
12. Florin NH, Harris AT (2007) Hydrogen production from biomass coupled with carbon dioxide capture: the implications of thermodynamic equilibrium. Int J Hydrogen Energy 32(17):4119–4134
13. Wang Y, Yoshikawa K, Namioka T, Hashimoto Y (2007) Performance optimization of two-staged gasification system for woody biomass. Fuel Process Technol 88(3):243–250
14. Wu C, Huang Q, Sui M, Yan Y, Wang F (2008) Hydrogen production via catalytic steam reforming of fast pyrolysis bio-oil in a two-stage fixed bed reactor system. Fuel Process Technol 89(12):1306–1316
15. Xu G, Murakami T, Suda T, Kusama S, Fujimori T (2005) Distinctive effects of CaO additive on atmospheric gasification of biomass at different temperatures. Ind Eng Chem Res 44(15):5864–5868
16. Yang H, Yan R, Chen H, Lee DH, Liang DT, Zheng C (2006) Pyrolysis of palm oil wastes for enhanced production of hydrogen rich gases. Fuel Process Technol 87(10):935–942
17. Demirbas A (2010) Hydrogen production from biomass via supercritical water gasification. Energy Sources, Part A: Recovery, Utilization Environ Effects 32(14):1342–1354
18. Demirbaş A (2005) Hydrogen production from biomass via supercritical water extraction. Energy Sources 27(15):1409–1417
19. Argun H, Kargi F, Kapdan IK, Oztekin R (2008) Biohydrogen production by dark fermentation of wheat powder solution: effects of C/N and C/P ratio on hydrogen yield and formation rate. Int J Hydrogen Energy 33(7):1813–1819
20. Hallenbeck PC, Benemann JR (2002) Biological hydrogen production; fundamentals and limiting processes. Int J Hydrogen Energy 27(11–12):1185–1193
21. Raheem A, Memon LA, Abbasi SA, Yap YHT, Danquah MK, Harun R (2017) Potential applications of nanotechnology in thermochemical conversion of microalgal biomass. In: Nanotechnology for bioenergy and biofuel production. Springer, Berlin, pp 91–116
22. Pugazhendhi A, Shobana S, Nguyen DD, Banu JR, Sivagurunathan P, Chang SW, Ponnusamy VK, Kumar G (2019) Application of nanotechnology (nanoparticles) in dark fermentative hydrogen production. Int J Hydrogen Energy 44(3):1431–1440. https://doi.org/10.1016/j.ijhydene.2018.11.114
23. Li J, Xiao B, Yan R, Liu J (2009) Development of a nano-Ni-La-Fe/Al2O3 catalyst to be used for syn-gas production and tar removal after biomass gasification. BioResources 4(4):1520–1535
24. Lakshmana Reddy N, Cheralathan KK, Durga Kumari V, Neppolian B, Muthukonda Venkatakrishnan S (2018) Photocatalytic reforming of biomass derived crude glycerol in water: a sustainable approach for improved hydrogen generation using Ni (OH) 2 decorated TiO2 nanotubes under solar light irradiation. ACS Sustain Chem Eng 6(3):3754–3764
25. de Vlieger DJM, Thakur DB, Lefferts L, Seshan K (2012) Carbon nanotubes: a promising catalyst support material for supercritical water gasification of biomass waste. Chem Cat Chem 4(12):2068–2074
26. Larimi A, Khorasheh F (2018) Renewable hydrogen production by ethylene glycol steam reforming over Al2O3 supported Ni-Pt bimetallic nano-catalysts. Renew Energy 128:188–199. https://doi.org/10.1016/j.renene.2018.05.070
27. Han H, Cui M, Wei L, Yang H, Shen J (2011) Enhancement effect of hematite nanoparticles on fermentative hydrogen production. Biores Technol 102(17):7903–7909
28. Burch R (2006) Gold catalysts for pure hydrogen production in the water–gas shift reaction: activity, structure and reaction mechanism. Phys Chem Chem Phys 8(47):5483–5500
29. Taherdanak M, Zilouei H, Karimi K (2016) The effects of Fe0 and Ni0 nanoparticles versus Fe2+ and Ni2+ ions on dark hydrogen fermentation. Int J Hydrogen Energy 41(1):167–173. https://doi.org/10.1016/j.ijhydene.2015.11.110

30. Nasr M, Tawfik A, Ookawara S, Suzuki M, Kumari S, Bux F (2015) Continuous biohydrogen production from starch wastewater via sequential dark-photo fermentation with emphasize on maghemite nanoparticles. J Ind Eng Chem 21:500–506. https://doi.org/10.1016/j.jiec.2014.03.011
31. Liu Z, Lv F, Zheng H, Zhang C, Wei F, Xing X-H (2012) Enhanced hydrogen production in a UASB reactor by retaining microbial consortium onto carbon nanotubes (CNTs). Int J Hydrogen Energy 37(14):10619–10626. https://doi.org/10.1016/j.ijhydene.2012.04.057
32. Beckers L, Hiligsmann S, Lambert SD, Heinrichs B, Thonart P (2013) Improving effect of metal and oxide nanoparticles encapsulated in porous silica on fermentative biohydrogen production by Clostridium butyricum. Biores Technol 133:109–117. https://doi.org/10.1016/j.biortech.2012.12.168
33. Zhang Y, Shen J (2007) Enhancement effect of gold nanoparticles on biohydrogen production from artificial wastewater. Int J Hydrogen Energy 32(1):17–23. https://doi.org/10.1016/j.ijhydene.2006.06.004
34. Aradi A, Roos J, Jao T-C (2013) Nanoparticle catalyst compounds and/or volatile organometallic compounds and method of using the same for biomass gasification. Google Patents,
35. Mullai P, Yogeswari MK, Sridevi K (2013) Optimisation and enhancement of biohydrogen production using nickel nanoparticles—a novel approach. Biores Technol 141:212–219. https://doi.org/10.1016/j.biortech.2013.03.082
36. Sınağ A, Yumak T, Balci V, Kruse A (2011) Catalytic hydrothermal conversion of cellulose over SnO2 and ZnO nanoparticle catalysts. J Supercrit Fluids 56(2):179–185. https://doi.org/10.1016/j.supflu.2011.01.002
37. Hsieh P-H, Lai Y-C, Chen K-Y, Hung C-H (2016) Explore the possible effect of TiO2 and magnetic hematite nanoparticle addition on biohydrogen production by Clostridium pasteurianum based on gene expression measurements. Int J Hydrogen Energy 41(46):21685–21691. https://doi.org/10.1016/j.ijhydene.2016.06.197
38. Li J, Yan R, Xiao B, Liang DT, Du L (2008) Development of nano-NiO/Al2O3 catalyst to be used for tar removal in biomass gasification. Environ Sci Technol 42(16):6224–6229. https://doi.org/10.1021/es800138r
39. Zhao W, Zhang Y, Du B, Wei D, Wei Q, Zhao Y (2013) Enhancement effect of silver nanoparticles on fermentative biohydrogen production using mixed bacteria. Biores Technol 142:240–245. https://doi.org/10.1016/j.biortech.2013.05.042
40. Zhao W, Zhao J, Chen GD, Feng R, Yang J, Zhao YF, Wei Q, Du B, Zhang YF (2011) Anaerobic biohydrogen production by the mixed culture with mesoporous Fe3O4 nanoparticles activation. Trans Tech Publ, pp 1528–1531

Chapter 6
Hydrogen from Miscellaneous Sources and Nanotechnology

6.1 Introduction

Apart from the sources mentioned in the previous chapters, there are several other important substances that generate H_2 upon chemical or physical treatment. For instance, when aluminum and NaOH are allowed to react they result in the generation of hydrogen. Al is the most abundantly found crustal metal. The metal can be fully recycled, and hence is also called as "viable metal." Al is a very lightweight metal which has several characteristics for being used in the generation of sustainable energy. It is also used in energy storage batteries. When employed with highly alkaline electrolyte the reduction potential of pure Al can be as low as -2.33 V relative to the standard hydrogen potential [1]. The formation of hydrogen in corrosion reaction of Al leads to the use of Al in generation of sustainable hydrogen. Aluminum-based generation of hydrogen requires 2% energy and the emission of carbon dioxide is 4% of the amount generated by the traditional methods of hydrogen generation. Al and Al-based alloys are excellent candidates of onboard vehicle hydrogen generation [2]. In strongly alkaline solution, the hydroxide ions (OH^-) are generated. These ions are capable of destroying the protective layer of the oxide formed on the surface aluminum, resulting in the formation of $AlO_2{}^-$. Because of this the Al and the Al-based alloys dissolved readily in the alkaline solutions. The reaction of the Al with alkalis occurs at room temperature resulting in the formation of hydrogen. In the chemical reactions below, it is shown that how Al reacts with NaOH (a strong alkali) for generation of hydrogen [3].

$$2Al + 6H_2O + 2NaOH \rightarrow 2NaAl(OH)_4 + 3H_2$$

$$NaAl(OH)_4 \rightarrow NaOH + Al(OH)$$

$$2Al + 6H_2O \rightarrow 2Al(OH)_3 + 3H_2$$

S. Farrukh et al., *Nanotechnology and the Generation of Sustainable Hydrogen*, Green Energy and Technology, https://doi.org/10.1007/978-3-030-60402-8_6

Municipal solid waste is also used as the source of hydrogen. Different components of the waste like biological substances, organic materials, and other undesirable components like plastic debris are involved in the generation of hydrogen. Different studies have reported the use of these materials collectively and individual for the preparation of hydrogen. In a study, post-consumer waste plastic obtained from the municipal solid waste was used for the hydrogen generation by catalytic steam reforming coupled with the gasification [4]. Waste plastic is one of the biggest environmental hazards of the modern age. Conversion of the plastic into hydrogen reduces the pollution on one hand and provides the clean energy on the other hand. Several studies have reported the use of the plastics for the generation of sustainable hydrogen. A study has reported the use of mechanochemical treatment for the generation of sustainable hydrogen from the waste plastic [5]. A study has reported the simultaneous conversion of plastics into hydrogen-rich syngas and the carbon nanotubes [6]. The waste plastic is commonly recycled chemically either by pyrolysis or by gasification or by both of them for the generation of important hydrocarbons. When the plastics are subjected to gasification, syngas is produced with large percentage of hydrogen. Additionally, the plastics comes with a higher hydrogen content and heating value, as compared to other types of municipal solid wastes and biomass, hence making them the most suitable candidate for hydrogen generation from the whole solid wastes [7].

Ammonia is another source of hydrogen generation. Ammonia and many of its derivatives are used as the potential source of sustainable hydrogen. Ammonia as the source of hydrogen seems appealing as it does not generate oxides of nitrogen or carbon (carbon dioxide or carbon monoxide) as the by-product, and also the concentration of the unreactive ammonia during the chemical reaction can be lowered to 200 ppb by employing appropriate adsorber [8]. It has been reported that the generation of H_2 by the decomposition of ammonia is economically more feasible as compared to the reformation of methanol [9]. The evolution of hydrogen from ammonia requires high temperature and suitable catalyst [10]. A chemical reaction for the generation of hydrogen from ammonia can be given as [11].

$$2NH_3 + 6OH^- \rightarrow N_2(g) + 6H_2O + 6e^-$$

$$6H_2O + 6e^- \rightarrow 3H_2(g) + 6OH^-$$

Oxidation from metal alloys of salt solution results in the generation of hydrogen. Magnesium alloys are notably important in this regard. It was found that the binary alloys of Mg are capable of reacting with solutions of sodium chloride in water [12]. Transition metals like Fe, Cu, Co, or Ni when mixed with Mg even in the small composition like 0.2 wt.% result in the improved oxidation ability of the magnesium. This can be associated to the high reduction potential of the transition metals [13]. A simple reaction of magnesium reacting with salts for the generation of hydrogen is given in the equation below [14]:

$$2NH_4Cl + Mg \rightarrow MgCl_2 + 2NH_3 + H_2$$

Similarly, different hydrides like sodium borohydride are also capable of generating the sustainable hydrogen. Hydrides of metals liberate hydrogen when decomposed thermally. A striking characteristic of hydrides is that they can be generated again in a reversible reaction. However, the hydrides have very small percentage of hydrogen production. The hydrides of magnesium-based alloys exhibit higher percentage of hydrogen production, at considerably high temperatures, hence making their use impractical for generation of hydrogen. Another major problem of the hydrides when used as the source of hydrogen is high air sensitivity. The general chemical reaction of hydrogen liberation from hydrolysis of hydrides can be understood by the following reaction using sodium borohydride ($NaBH_4$) as the hydrogen source:

$$NaBH_4 + H_2O \rightarrow NaBO_2 + 4H_2$$

$NaBH_4$ is one of the important hydrides of boron. The compound is stable in dry air and is easy to handle as compared to the other chemicals of such nature. Hydrogen can be generated by the action of catalysts or by hydrolysis of sodium borohydride. Sodium borohydride is commonly employed hydride for the generation of hydrogen because of its several advantages over the others such as the solutions of the compound are non-flammable; the chemical reactions liberate environmentally benign products, the reactions of sodium borohydride allow the rate control of H_2 production, the by-product of the reaction $NaBO_2$ can be easily recycled, and the reactions occur at comparatively low temperatures. In the hydrolysis of the sodium borohydride, the amount of hydrogen released is 10.8 wt.%, where half of the hydrogen comes from the water [15]. Different catalysts have been reported for the generation of hydrogen with $NaBH_4$. $Ru(acac)_3$ was used as the catalyst for hydrolysis of $NaBH_4$, *Keceli* et al. [16]. Silica-supported ferromagnetic compounds have also been found to assist the hydrolysis of the compound as catalysts [17]. Transition metals like Ni, Pt/C, and Ru/C also serve as the important catalysts of the reaction [18–20]. Other than $NaBH_4$ important hydrides like $LiBH_4$ and KBH_4 are also employed for the generation of hydrogen [21].

Hydrous hydrazine is also used as the raw material for the generation of hydrogen. It is a derivative of ammonia. Hydrous hydrazine is a compound in liquid state at room temperature. It has high contents of hydrogen. About 8.0 wt.% of hydrogen can be obtained from the decomposition of hydrazine at mild conditions of reaction. The compound only generates nitrogen as the by-product on complete dissociation. The chemical equation for the reaction can be given as [22].

$$N_2H_4 \rightarrow N_2 + 2H_2$$

The use of the suitable catalyst is an absolute necessity for obtaining hydrogen form hydrazine because the reaction is capable of proceeding in more than one pathway. For instance, if the decomposition is not controlled by appropriate catalysts it can proceed in the following manner [23]:

$$N_2H_4 \rightarrow 4NH_3 + N_2$$

Various metallic catalysts are used for the decomposition of the hydrazine for the generation of hydrogen.

Several of abovementioned processes involve the extensive use of nanotechnology in the liberation of hydrogen. This chapter will explain the involvement of several nanomaterials in generation of hydrogen from miscellaneous sources.

6.2 Nano Hydride-Based Hydrogen Generation

Hydrogen can be obtained from the wide variety of the hydrides upon hydrolysis over certain types of catalysts. A study has reported the use of nanosized platinum dispersed on $LiCoO_2$ for the generation of hydrogen from the solution of lithium borohydride. The substrate ($LiAlH_4$) liberated stoichiometric amount of hydrogen over the prepared nanocatalysts. 8.6 wt.% of the hydrogen is generated when the amount of water is considered in the calculations. The chemical equation for the reaction can be given as.

$$LiBH_4 + 4H_2O \rightarrow LiBO_2.2H_2O + 4H_2$$

The said nanocatalysts were prepared by calcining the lithium cobalt powder 250 °C for 5 h, followed by the coating of Pt by employing the oxide of the metal. The said materials were characterized with wide angle X-ray diffraction (XRD), transmission electron microscopy (TEM), and energy-dispersive X-ray spectroscopy [24].

Bimetallic magnetic nanoparticles of Ni and Ru constrained on the resin beads were fabricated and used as the catalysts for production of hydrogen from alkaline solutions of sodium borohydride. The said nanomaterials were fabricated with chemical reduction and electroless deposition processes. For the synthesis of the nanomaterials, resin beads were immersed into the $RuCl_3$ solution at the room temperature so that the chelating reaction can occur. The system was allowed to stir for 1 h. Afterward, the Ru-chelated beads were washed with deionized water, so as to remove the excessive product. The product was then heated at 70 °C and followed by the reduction of Ru^{3+} to Ru. In this way, electroless deposition of Ni became possible which is basically an autocatalytic reaction and only occurs on the conductive metal surfaces. The prepared nanomaterials were analyzed with scanning electron microscopy (SEM), EDS, X-ray photoelectron spectroscopy (XPS), and magnetometer. The nanocatalysts can be recovered with spent solution of $NaBH_4$ with permanent magnets. This is due to the inherent soft ferromagnetic nature of the prepared nanomaterial. The recovery of the catalyst leads to the cost reduction of process. 400 mL/min/g of hydrogen was achieved from 10 wt.% solution of $NaBH_4$ with 5 wt. % NaOH at 35 °C 5 wt. % NaOH using the fabricated nanocatalysts [25].

As mentioned earlier, Al and the Al-based alloys are very significant in the generation of hydrogen. Nanocrystalline Al hydrides also known as alanates are very significant hydrogen storage substances for proton exchange membrane fuel cell applications. In order to use the alanates for the hydrogen storage or generation applications, several factors like amount of reversible hydrogen, thermodynamic and kinetic parameters have to be optimized. Ti catalysts have seen to improve the uptake and release of kinetics of hydrogen. In an investigation, different nanocrystalline Al hydrides like magnesium alanate $Mg\ (AlH_4)_2$ have been synthesized which contains 9.3 wt.% calculated amount of hydrogen. The fabricated alanate when allowed to decompose at ~160 °C, the three-fourth of the hydrogen is generated, and the last part of the hydrogen can also be obtained at temperatures as high as 300 °C, which is not suitable for proton exchange membrane fuel purposes. The chemical reaction for the process can be given as.

$$MgH_2 \rightarrow Mg + H_2$$

Afterward, the Mg reacts with Al to give the following reaction:

$$Mg + 2Al \rightarrow 12Al_3Mg_2 + 12Al$$

The formation of the intermetallic compound is confirmed by the XRD analysis. The equations give the rough estimation of the decomposition of the material [26].

Nanoclusters of intrazeolite cobalt were used in an investigation for the hydrogen liberation from sodium borohydride. The nanoclusters were fabricated via the ion-exchange method. In practice, the Na^+ ions of the zeolite were exchanged by the Co^{2+} ions, followed by the subsequent reduction (of Co^{2+} ions inside the zeolite-Y) with $NaBH_4$ as imitator at room temperature. The product thus formed was separated as solid material. The formed product was analyzed with XRD, inductively coupled plasma optical emission spectrometry (ICP-OES), SEM, TEM, XPS, Raman spectroscopy, and N_2 adsorption technique. The nanoclusters were evaluated for their catalytic activity in the hydrolysis of sodium borohydride solution with and without the base. The nanoclusters have depicted better results in the basic solution. The study reported the record production turnover of hydrogen from the hydrolysis of the sodium borohydride in alkaline solution [27].

Nurettin et al. have reported the use of ruthenium and magnetic iron nanoparticles as the catalyst for the hydrolysis of $NaBH_4$. The nanoparticles in the study were used in the form of composites on the hydrogel matrix. The hydrogels were prepared from -acrylamido-2-methyl-1-propansulfonic acid, N, N′-methylenebisacrylamide (MBA), and 2,2′-azobis(2-methylpropionamidine) dihydrochloride as monomer, crosslinker, and UV initiator, respectively. The resulting hydrogel is pH sensitive, hydrophilic, and have the ability to absorb the metal ions. The Ru and Fe nanoparticles inside the hydrogels were prepared separately. The prepared product was characterized with TEM. The study found have that the repetitive use of the prepared material resulting in the leeching of the metal particles; however, they can be loaded again, and maximum efficiency can be achieved. The catalysts were later on used to reduce the $NaBH_4$ in the alkaline environment. Chemical equation for the reaction can be given as [28].

$$NaBH_4 + H_2O \rightarrow NaBO_2 + 4H_2$$

Similar studies with same hydrogel matrix containing cobalt and nickel nanoparticles have also been reported for the hydrolysis of alkaline sodium borohydride solution for the generation of sustainable hydrogen [29, 30].

6.3 Waste Material-Based Hydrogen Generation

Different waste materials that are involved in the generation of hydrogen require extensive use of nanotechnology. In case of the plastics, nanomaterials are involved in different ways. In some cases, the nanomaterials are produced simultaneously from the polymeric waste plastic along with sustainable hydrogen (normally carbon nanotubes are obtained by the process). However, some studies have reported the recovery of metallic nanoparticles from the plastics but no hydrogen was obtained along with them [31]. Some studies have reported the formation of particular nanoparticles from different types of waste plastics which are involved in the generation of hydrogen. Some of these works will be discussed here.

A study published in 2012 reported the generation of simultaneous generation of hydrogen-rich synthesis gas and supreme quality multiwalled carbon nanotubes from the waste plastics. The conversion was achieved through gasification instead of thermal recycling. The process of the conversion was designed in a manner so that it can be altered so as to achieve desired end product. In the practice, the waste polypropylene was used as the waste plastic and the steam gasification of the substrate was achieved by Ni/Zn–Al and Ni/Ca–Al catalysts in a two-step reaction. The carbon nanotubes were obtained in the form of deposition on the surface of the catalysts and were characterized with Raman spectroscopy, SEM, and TEM. The mass of the obtained nanotubes was determined through temperature-programmed oxidation (TPO) of the catalyst with product deposition. The percentage yield of sustainable hydrogen obtained through the steam gasification of the polymer was found to be 34.1 vol. %. The study proposed 16,800 Nm^3 per day generation of the hydrogen by the process if the plants are installed with 10 tons per day capacity of the plastic [6]. A graphical explanation of plastic conversion into carbon nanotubes and sustainable hydrogen is given in Fig. 6.1.

In another study, novel waste polyethylene-based nanomaterials were created and were employed in the generation of hydrogen from electrochemical reactions. These nanomaterials involve carbon nanotubes decorated with Ni/Pd nanoparticles. In practice, the nanomaterial was prepared by disintegrating the polyethylene waste plastic in supercritical CO_2 at high temperatures. This led to the formation of a mixture

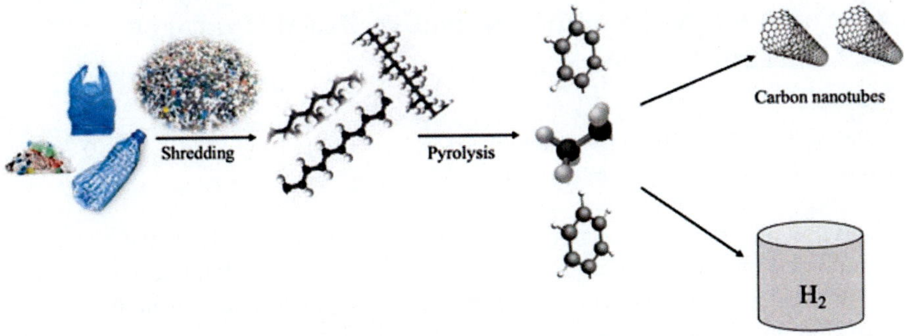

Fig. 6.1 Conversion of plastic in carbon nanotubes and sustainable hydrogen

aromatic hydrocarbons, CO_2 and CO. Supercritical CO_2 ensured the complete and quick dissociation. In a parallel reaction, Ni acetate tetrahydrate was also pyrolyzed so as to get the Ni nanoparticles. Ni nanoparticles are excellent catalyst for the growth of carbon nanotubes. The nanotubes were then allowed to grow by tip-growth model via the Ni nanoparticle-supported small organic molecule thermal decomposition. The Pd nanoparticles were allowed to develop on the surface of the nanoparticles by using the aqueous solution of $PdCl_2$ (liquid phase synthesis followed by the subsequent reduction of the adsorbed Pd^{2+} ions. The prepared nanomaterials were employed in electrochemical hydrogen evolution reaction via the splitting of water. The carbon nanotube performance was poor when employed as catalyst; however, it improved drastically when decorated with the Pd nanoparticles [32]. This can be associated to the high reduction potential of the Pd.

A group of researchers have reported a work which addresses the simultaneous production of carbon-black nanospheres and hydrogen from the thermal pyrolysis of different waste polymers. For the purpose, the samples of the four different waste plastic polymers, i.e., polyvinyl chloride, polypropylene, acrylonitrile butadiene styrene, and polyethylene, were charged into a furnace with twin thermal plasma jet. This led to the pyrolysis of the substrate polymers. The pyrolysis resulted in the formation of different gases and residues. The formed gaseous components were analyzed with optical emission spectrometer (OES) and gas chromatography [33] for the determination of the composition of the components. The analysis indicated that the gaseous product contains H_2, CH_4, C_2H_2, C_2H_6, C_2H_4, C_3H_8, C_3H_6, C_4H_{10}, C_5H_{12}, and C_4. Analysis of all four polymers gaseous products showed that the hydrogen has maximum concentration among all other gases. Another important fact revealed by the analysis is that none of the gaseous mixture showed the presence of carbon monoxide. The formed solid residue was analyzed with XRD, TEM, and SEM. The characterization showed the ultrapure carbon-black nanospheres were formed by the acrylonitrile butadiene styrene pyrolytic decomposition [34].

6.4 Ammonia/Ammonium-Derivative-Based Hydrogen Generation

As stated, earlier ammonia and its derivative are important source of hydrogen liberation. Some of the ammonia derivatives employ nanotechnology for hydrogen generation.

Serdar et al. have used ceria-supported rhodium nanomaterials for the hydrolysis of ammonia borane to liberate hydrogen. The researchers elaborated the effect of different oxides on the catalytic ability of rhodium nanoparticles during the hydrolysis of ammonia borane for hydrogen generation. Different oxide supports like CeO_2, SiO_2, Al_2O_3, TiO_2, ZrO_2, and HfO_2 were used. However, ceria offered the best catalytic with rhodium nanoparticles for ammonia borane hydrolysis. The said nanomaterial consisted of Rh(0) nanoparticles restrained on nanoceria, i.e., Rh(0)/CeO_2. The nanomaterials were synthesized via impregnation process. In practice, the Rh(III) ions were allowed to deposit on the ceria surface by using $RhCl_3 \cdot 3H_2O$, and the deposition was followed by the reduction of ions to Rh(0) with sodium borohydride at room temperature. The nanomaterials thus obtained were separated via centrifugation and analyzed with inductively coupled plasma optical emission spectroscopy (ICP-OES), TEM, SEM, and XPS. The catalytic ability of the prepared nanomaterials was determined at different wt.% loading of the Rh nanoparticles in nanoceria for the hydrolysis of ammonia borane. The catalysts depicted best activity with 0.1 wt.% loading of Rh nanoparticles. The nanoparticles showed record turnover frequency value of 2010/min at ~25 °C. The remarkable catalytic activity of the material is associated to the reducible nature of nanoceria. The prepared nanomaterials are reusable catalysts with the ability to retain 67% initial activity even in the fifth cycle [35].

Another investigation has reported the use of Pd nanoparticles maintained on cobalt ferrite (Pd(0)/$CoFe_2O_4$) as a catalyst for the hydrogen generation via hydrolysis of ammonia borane. The catalyst depicted turnover frequency of 290/min at 25 °C. The stability of the substance was enhanced with the use of polydopamine-covered cobalt ferrite in place of exposed cobalt ferrite. The coating of polydopamine was obtained via pH-dependent self-polymerization with dopamine hydrochloride at 25 °C. The Pd nanoparticles were loaded on the support material (polydopamine-coated cobalt ferrite) via two-step process involving impregnation and reduction. The materials were characterized with Fourier transform infrared (FTIR), XRD, attenuated total reflectance [36], TEM, SEM, EDS, ICP-OES, and Brunauer–Emmett–Teller (BET). The analyzation revealed that Pd nanoparticles have average size of ~1.4 nm and have complete dispersion on polydopamine layer, where the thickness of the layer was found to be ~ 8.6 nm. The value of turnover frequency was determined to 175/min at ~25.0 °C. The chemical equation for the reaction can be given as [37].

$$NBH_6 + 2H_2O \rightarrow NH^{4+} + BO^{2-} + 3H_2$$

Production of hydrogen from hydrazine also involves the extensive use of nanotechnology. Different studies have reported the use of various nanomaterials in hydrogen generation from hydrazine. Modified silver nanoparticles designated as C60(Ag@C60) are regarded as remarkable electrocatalysts for hydrazine oxidation to generate hydrogen [38]. Similarly, titanate nanotubes enclosing Ni nanoparticles have the ability to catalyze selective oxidation of hydrazine for hydrogen generation [39].

6.5 Conclusion

There are several miscellaneous sources which contribute toward the generation of hydrogen in varying amounts. As proven from the above discussion, nanotechnology is paving the path for easing the generation of H_2, both in terms of time and cost from these miscellaneous sources. Efforts must be made for the use of the miscellaneous sources as portable mediums that can be employed in batteries and other small devices for the generation of hydrogen.

References

1. Rand DAJ, Woods R (1981) Batteries for electric vehicles
2. Schlapbach L, Züttel A (2011) Hydrogen-storage materials for mobile applications. In: Materials for sustainable energy: a collection of peer-reviewed research and review articles from nature publishing group. World Scientific, pp 265–270
3. Wang HZ, Leung DYC, Leung MKH, Ni M (2009) A review on hydrogen production using aluminum and aluminum alloys. Renew Sustain Energy Rev 13(4):845–853. https://doi.org/10.1016/j.rser.2008.02.009
4. Wu C, Williams PT (2010) Pyrolysis–gasification of post-consumer municipal solid plastic waste for hydrogen production. Int J Hydrogen Energy 35(3):949–957. https://doi.org/10.1016/j.ijhydene.2009.11.045
5. Qiwu Z, Fumio S (2009) Generation of hydrogen and syngas from waste plastics and wood-chip with an aid of mechanochemical treatment. J Soc Powder Technol Jpn 46(6):484–488
6. Wu C, Wang Z, Wang L, Williams PT, Huang J (2012) Sustainable processing of waste plastics to produce high yield hydrogen-rich synthesis gas and high quality carbon nanotubes. RSC Adv 2(10):4045–4047
7. Ahmed II, Gupta AK (2009) Hydrogen production from polystyrene pyrolysis and gasification: characteristics and kinetics. Int J Hydrogen Energy 34(15):6253–6264
8. Chellappa AS, Fischer CM, Thomson WJ (2002) Ammonia decomposition kinetics over Ni-Pt/Al2O3 for PEM fuel cell applications. Appl Catal a 227(1–2):231–240
9. Metkemeijer R, Achard P (1994) Ammonia as a feedstock for a hydrogen fuel cell; reformer and fuel cell behaviour. J Power Sources 49(1–3):271–282
10. Yin S-F, Zhang Q-H, Xu B-Q, Zhu W-X, Ng C-F, Au C-T (2004) Investigation on the catalysis of COx-free hydrogen generation from ammonia. J Catal 224(2):384–396
11. Vitse F, Cooper M, Botte GG (2005) On the use of ammonia electrolysis for hydrogen production. J Power Sources 142(1–2):18–26

12. Kravchenko OV, Sevastyanova LG, Urvanov SA, Bulychev BM (2014) Formation of hydrogen from oxidation of Mg, Mg alloys and mixture with Ni Co, Cu and Fe in aqueous salt solutions. Int J Hydrogen Energy 39(11):5522–5527. https://doi.org/10.1016/j.ijhydene.2014.01.181
13. Ferrando WA (1989) Review of corrosion and corrosion control of magnesium alloys and composites. J Mater Eng 11(4):299–313
14. Tikhonov VN (1973) Analytical chemistry of magnesium. In: Analytical chemistry of elements. Nauka Publisher Moscow, p 10
15. Liu BH, Li ZP (2009) A review: Hydrogen generation from borohydride hydrolysis reaction. J Power Sources 187(2):527–534. https://doi.org/10.1016/j.jpowsour.2008.11.032
16. Keçeli E, Özkar S (2008) Ruthenium (III) acetylacetonate: a homogeneous catalyst in the hydrolysis of sodium borohydride. J Mol Catal Chem 286(1–2):87–91
17. Shih Y-J, Su C-C, Huang Y-H, Lu M-C (2013) SiO2-supported ferromagnetic catalysts for hydrogen generation from alkaline NaBH4 (sodium borohydride) solution. Energy 54:263–270
18. Guella G, Patton B, Miotello A (2007) Kinetic features of the platinum catalyzed hydrolysis of sodium borohydride from 11B NMR measurements. J Phys Chem C 111(50):18744–18750
19. Metin Ö, Özkar S (2007) Hydrogen generation from the hydrolysis of sodium borohydride by using water dispersible, hydrogenphosphate-stabilized nickel (0) nanoclusters as catalyst. Int J Hydrogen Energy 32(12):1707–1715
20. Shang Y, Chen R (2006) Hydrogen storage via the hydrolysis of NaBH4 basic solution: optimization of NaBH4 concentration. Energy Fuels 20(5):2142–2148
21. Şahin Ö, Dolaş H, Özdemir M (2007) The effect of various factors on the hydrogen generation by hydrolysis reaction of potassium borohydride. Int J Hydrogen Energy 32(13):2330–2336
22. Singh SK, Zhang X-B, Xu Q (2009) Room-temperature hydrogen generation from hydrous hydrazine for chemical hydrogen storage. J Am Chem Soc 131(29):9894–9895
23. Singh SK, Iizuka Y, Xu Q (2011) Nickel-palladium nanoparticle catalyzed hydrogen generation from hydrous hydrazine for chemical hydrogen storage. Int J Hydrogen Energy 36(18):11794–11801. https://doi.org/10.1016/j.ijhydene.2011.06.069
24. Kojima Y, Suzuki K-i, Kawai Y (2006) Hydrogen generation from lithium borohydride solution over nano-sized platinum dispersed on LiCoO2. J Power Sources 155(2):325–328. https://doi.org/10.1016/j.jpowsour.2005.04.019
25. Liu C-H, Chen B-H, Hsueh C-L, Ku J-R, Jeng M-S, Tsau F (2009) Hydrogen generation from hydrolysis of sodium borohydride using Ni–Ru nanocomposite as catalysts. Int J Hydrogen Energy 34(5):2153–2163
26. Fichtner M, Engel J, Fuhr O, Kircher O, Rubner O (2004) Nanocrystalline aluminium hydrides for hydrogen storage. Mater Sci Eng, B 108(1):42–47. https://doi.org/10.1016/j.mseb.2003.10.036
27. Rakap M, Özkar S (2009) Intrazeolite cobalt (0) nanoclusters as low-cost and reusable catalyst for hydrogen generation from the hydrolysis of sodium borohydride. Appl Catal B 91(1–2):21–29
28. Sahiner N, Ozay O, Inger E, Aktas N (2011a) Controllable hydrogen generation by use smart hydrogel reactor containing Ru nano catalyst and magnetic iron nanoparticles. J Power Sources 196(23):10105–10111. https://doi.org/10.1016/j.jpowsour.2011.08.068
29. Ozay O, Aktas N, Inger E, Sahiner N (2011) Hydrogel assisted nickel nanoparticle synthesis and their use in hydrogen production from sodium boron hydride. Int J Hydrogen Energy 36(3):1998–2006
30. Sahiner N, Ozay O, Inger E, Aktas N (2011b) Superabsorbent hydrogels for cobalt nanoparticle synthesis and hydrogen production from hydrolysis of sodium boron hydride. Appl Catal B 102(1–2):201–206
31. Dai W, Lu L, Han Y, Wang L, Wang J, Hu J, Ma C, Zhang K, Mei T (2019) Facile Synthesis of Mo2C Nanoparticles from Waste Polyvinyl Chloride. ACS Omega 4(3):4896–4900. https://doi.org/10.1021/acsomega.8b02856
32. Gao L, Zhou F, Chen Q, Duan G (2018) Generation of Pd@ Ni-CNTs from polyethylene wastes and their application in the electrochemical hydrogen evolution reaction. Chem Sel 3(19):5321–5325

33. Okan BS, Menceloğlu Y, Ozunlu BG, Yagci YE (2020) Graphene from waste tire by recycling technique for cost-effective and light-weight automotive plastic part production. AIP Publishing LLC, p 020046
34. Mohsenian S, Esmaili MS, Fathi J, Shokri B (2016) Hydrogen and carbon black nano-spheres production via thermal plasma pyrolysis of polymers. Int J Hydrogen Energy 41(38):16656–16663. https://doi.org/10.1016/j.ijhydene.2016.05.150
35. Akbayrak S, Tonbul Y, Özkar S (2016) Ceria supported rhodium nanoparticles: Superb catalytic activity in hydrogen generation from the hydrolysis of ammonia borane. Appl Catal B 198:162–170. https://doi.org/10.1016/j.apcatb.2016.05.061
36. Hallenbeck PC, Benemann JR (2002) Biological hydrogen production; fundamentals and limiting processes. Int J Hydrogen Energy 27(11–12):1185–1193
37. Manna J, Akbayrak S, Özkar S (2017) Palladium(0) nanoparticles supported on polydopamine coated CoFe2O4 as highly active, magnetically isolable and reusable catalyst for hydrogen generation from the hydrolysis of ammonia borane. Appl Catal B 208:104–115. https://doi.org/10.1016/j.apcatb.2017.02.037
38. Narwade SS, Mulik BB, Mali SM, Sathe BR (2017) Silver nanoparticles sensitized C60(Ag@C60) as efficient electrocatalysts for hydrazine oxidation: Implication for hydrogen generation reaction. Appl Surf Sci 396:939–944. https://doi.org/10.1016/j.apsusc.2016.11.065
39. Wang H, Wu L, Jia A, Li X, Shi Z, Duan M, Wang Y (2018) Ni nanoparticles encapsulated in the channel of titanate nanotubes: Efficient noble-metal-free catalysts for selective hydrogen generation from hydrous hydrazine. Chem Eng J 332:637–646. https://doi.org/10.1016/j.cej.2017.09.126

Chapter 7
Physisorption

7.1 Introduction

One of the major reasons of why hydrogen is still not used as the primary source fuel is the limitations in storage of hydrogen. The storage of the hydrogen should be reversible at ambient conditions or it can be released from the storage material on demand. Physisorption can be described as non-dissociative surface association of the hydrogen gas with the solid surfaces. The interactions involve weak van der Waals associations between the gaseous hydrogen molecules and the adsorbent [1].

The adsorption of hydrogen either chemical or physical involves varying energy changes. A simplified version of these energies is shown in Fig. 7.1 chemically or physically bonded to the surface. When the hydrogen gas comes in the vicinity of the hydrogen-adsorbing material, some of the approaching molecules (depending upon the nature of the adsorbent) develop physical association or weak van der Waals interaction with the material surface and get physiosorbed on it. When sufficiently high energy is provided either by altering the temperature or pressure, the molecules of hydrogen dissociate into atoms and form chemical bonds with the adsorbent, i.e., chemisorption. Reversing the conditions of temperature or pressure releases the adsorbed hydrogen. There is a large array of substances that are involved in hydrogen physisorption. The specific surface area and bonding energy (expressed in m^2g^{-1}) are the major performance indicators of any material to be used for physisorption of hydrogen [2].

Different carbon materials like carbons nanotubes (both single walled and multi-walled), zeolites, activated carbons, COFs, fullerene (nanocages), and covalent-organic frameworks are capable of storing hydrogen by physisorption [3–5]. This ability of the carbon nanomaterials can be attributed to their sufficient micro-porosity, large surface areas ranging from 1000 to 3000 m^2g^{-1}, low mass, and fine adsorption ability. However, at ambient conditions, small amounts of hydrogen (1–3% maximum) get weakly physiosorbed on carbon surfaces [6]. Better outcomes can be achieved by lowering of temperature and pressure. Liquid nitrogen (−196 °C)

S. Farrukh et al., *Nanotechnology and the Generation of Sustainable Hydrogen*,
Green Energy and Technology, https://doi.org/10.1007/978-3-030-60402-8_7

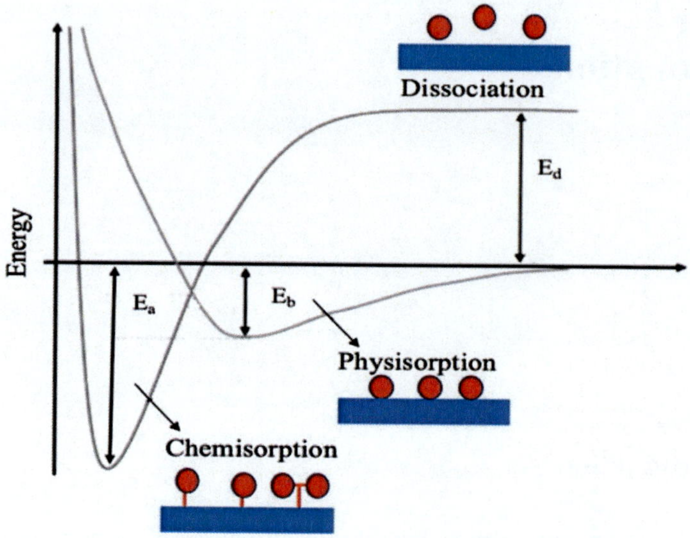

Fig. 7.1 Energy distribution diagram of hydrogen adsorption on solid material, where E_a = activation energy required for chemisorption, E_b = physisorption energy, E_d = dissociation energy

is commonly employed for standard low-temperature value and 20 bars hydrogen pressure is used. In these conditions, with large surface area carbons (~3000 m^2g^{-1}) ~5 wt.% gravimetric densities of hydrogen physisorption have been reported [7].

Metal organic frameworks (MOFs) have also been employed as the potential hydrogen storage material. MOFs are extremely porous organometallic hybrid structure. MOFs are nanoporous substances with high specific area and low density. They are formed by the linking of metal oxide species with the organic moieties [8]. MOFs are actively involved in the storage of hydrogen. Promising storage capabilities, greater than 8 wt.%, have been shown by nanoporous metal benzenedicarboxylate at −196 °C and 1.6 MPa [9].

Many physisorbing substances have noteworthy gravimetric densities and sufficient adsorption and desorption kinetics but they need cryogenic temperatures for proper functioning. Need of the hour is to prepare new substances which are capable of hydrogen storage applications at ambient temperature while keeping other important characteristics required for modern storage applications of the gas [2]. It should be mentioned that physisorption singly is not enough for the development of hydrogen storage containers, additional developments are needed in this area. Nanomaterials are actively involved in the physisorption of hydrogen. Some of the important nanomaterials with extensive use in physisorption of hydrogen are discusses here.

7.2 Hydrogen Physisorption and Fullerenes

Fullerenes are nanomaterials with sp^2 hybridized hollow carbon cages. Commonly found as C_{60}, but other order materials like C_{70}, C_{78}, C_{28}, and C_{36} are also known [10]. Fullerenes are actively involved in the physisorption of hydrogen.

ZhiGang et al. have reported the barrierless physisorption to chemisorption of hydrogen molecules on the fullerenes doped with lightweight elements. The study employed local density approximation (LDA) within density functional theory (DFT) as process. The DFT process is capable of large systems at low computational costs. In practice, $C_{35}B$ fullerene brought onto the space of $C_{35}B–H_2$ stable system. The two $C_{35}B$ molecules now behave like forceps with each attaching itself to the opposite side of the small H_2 molecules resulting in its dissociation to H, H (with one H with each of $C_{35}B$). This is barrierless physisorption to chemisorption. The study suggested that H–H bond breaking occurs because of energy discharge as $C_{35}B$ gets near to the $C_{35}BH_2$. The evolved energy increased the bond length between H–H ultimately leading to chemisorption of H_2 molecules in the hydrogen storage method. Overall the energy decreases monotonously during this procedure [11].

In an interesting investigation, fullerene nanocages were filled with hydrogen to determine the hydrogen storage capacity of endohedral fullerenes. The nanocages of general formula $H_n@C_k$ were analyzed for their hydrogen-storing capacities via DFT. When large amount of hydrogen molecules were filled inside the nanocages, a few of them get chemisorbed with carbon atoms on the interior of the nanostructures. At most, 58 hydrogen atoms were found in interior of C60 nanocage in metastable structure. The breaking mechanism of the fullerene was studied employing ab initio molecular dynamics simulations. The pressure of hydrogen in the nanocage was assessed and it was found that this pressure is few times less than hydrogen metallization pressure. The study also developed the effect of hydrogen on C–C bond stretching in fullerene nanocages of random radii. This can help in assessing the amount of hydrogen that can be stored in larger radii fullerenes [4].

In an investigation, hydrogen storage capability of 3D periodic fullerene pillared graphene nanocomposites was studied for their hydrogen storage ability. The 3D nanomaterials consisted of fullerene entities covalently combined between the layers of graphene. Fullerenes were used as pillars in between the consecutive layers of the graphene. They were employed for adjusting the porosity and improving the hydrogen storage capabilities of the projected nanostructures. The volumetric and gravimetric hydrogen storage of the nanomaterials was studied by employing Monte Carlo calculations at both high and low pressures varying from 0.01 to 100 bars and at different temperatures, i.e., −196 °C and 25 °C. The results of simulations depicted that substantial improvement in hydrogen adsorption ability of the nanomaterials can be achieved with fitting assortment of fullerene size and the conditions of hydrogen loading. The calculation also showed that the said nanomaterials can store up to 10.3 wt.% hydrogen at −196 °C. Additionally, the nanomaterial capacity of releasing hydrogen surpassed 7.8 wt.% for the charge at −196 °C and 100 bar pressure. The discharge of the gas was achieved at −113 °C, 5 bar pressure [12].

Another investigation employs first-principle calculation within DFT for studying the hydrogen storage capability in Jahn–Teller slanted fullerenes modified with Ti. It is observed that that Ti atoms make two hexagonal pyramidal structures because of their high cohesive energy. Each Ti atom adsorbed four hydrogen molecules via Kubas interactions, with 0.33–0.76 eV adsorption energy per molecule of hydrogen. The calculations made in the study and the van't Hoff desorption temperature depicted that molecules of hydrogen are reversibly adsorbed under feasible thermodynamic conditions with 10.5 wt. % of hydrogen [13].

In an investigation, a novel carbon-based nanomaterial was analyzed for the hydrogen adsorption capacities. The nanoporous material can be described as the small fullerene units covalently enclosed by parallel layers of graphene in the form of sandwich. The fabricated nanomaterials have high surface-to-weight ratios and promising structural stability due to the mesoporous and microporous morphology of the nanostructures. The 3D nanomaterial model was prepared by fusing the small fullerene entities between the consecutive layers of graphene and then by stacking up the fullerene infused layers of graphene upon each other. The heat welding process run by molecular dynamic simulations was used for the preparation of the materials. Grand canonical Monte Carlo calculations were used for the determination of hydrogen adsorption abilities of the prepared nanostructures. In simulations, distinctive fullerene entities such as C_{180}, C_{320}, and C_{540} were regarded as the core of sandwich. The effects of lithium incorporation to the nanomaterials on hydrogen adsorption were also studied during simulations. The simulation analysis depicted sandwiched like nanostructure doped with lithium in Li-to-carbon ratio of 1 to 8 can surpass the gravimetric hydrogen storage capability of 5% whereas the undoped samples can attain 3.83% at −193 °C temperature and 1 bar pressure [14].

7.3 Hydrogen Physisorption and Carbon Nanotubes

Different material nanotubes are involved in physisorption of hydrogen; however, carbon nanotube is one of the best materials for the purpose. This is associated with large surface area, adjustable features, and lower mass density of the carbon nanotubes [15]. Here involvement of carbon nanotubes as well as other nanotubes in physisorption of hydrogen will be discussed.

Another investigation discovered the hydrogen adsorption ability of boron nanotubes and boron sheets via DFT calculations. The study also explained the electronic structure and geometry of both the boron nanotubes and sheets. The materials possessed discontinuous ups and downs in the layers of boron atoms forming buckled surfaces. The buckled surfaces have the height of ~ 0.8 Å, with 0.20 eV/atom more stabilization as compared to the corresponding flat materials. The ups and downs are not present in all of the boron nanotubes because the nanotubes with helicity do not allow this specific arrangement of the atoms. Despite having different geometries and the bonding features, both flat and buckled nanostructures are of metallic nature. The nanotubes depicted 30–60 meV/molecule H_2 physisorption energies on boron nanotubes and sheets. These energies are less than the energies exhibited by carbon nanotubes and graphene. The study found the 1 eV/molecule energy barrier from molecular physisorption to dissociative chemisorption of H_2 on these materials [16].

Different studies have reported different concentrations of hydrogen adsorption on carbon nanotubes. In an investigation, purified and modified samples of carbon nanotubes have been reported for up to 8 wt.% storage of hydrogen. Temperature-programmed desorption (TPD) spectroscopy was used for the analysis of adsorption of H_2 in carbon nanotubes. The characterization depicted that the hydrogen gets adsorbed in the empty cavities and canals of the carbon nanotubes. The same study has also envisioned that the carbon nanotubes with length and diameter of 2 and 1.63 nm, respectively, may achieve the target set by the USA Department of Energy [17].

Ye et al. have reported the physisorption of hydrogen on crystalline ropes on single-walled carbon nanotubes, with the hydrogen-storing capacity of more than 8 wt. %. The study elaborated that the H_2 first gets adsorbed at the exterior surfaces of the carbon nanotubes that are exposed. However, when the conditions of pressure and temperature are altered to 40 bar and −193 °C, a phase transition takes place and all the nanotubes get separated from each other and all their surfaces become exposed, allowing hydrogen to physiosorbed on them. A nanotube–nanotube cohesive energy is obtained due to the pressure of phase transition for most of the material of 5 meV per carbon atom. The crystalline order in the nanotubes has strong effect on this small cohesive energy. It was reported in the investigation that the H to C atoms ratio of about 1.0 was achieved for carbon nanotubes at −193 °C and > 12 MPa pressure [18].

Different types of carbon nanofibers were analyzed by *Chambers* et al. for the adsorption of hydrogen. They reported that the tubular nanofibers can store 11 wt.%, platelet form can adsorb 45 wt.%, and the herringbone can store 67 wt. % of hydrogen at 25 °C and 12 MPa pressure. The release of the hydrogen from these materials can be achieved by successive lowering of the pressure to approximate atmospheric conditions at 25 °C [19].

In a study, molecular dynamics simulation was used for physisorption of molecular hydrogen on single-walled vacant defect carbon nanotubes. The study focused on the effects of different parameters like size of vacant defect, pressure, and temperature on hydrogen physisorption. It was found that hydrogen can be stored inside the carbon nanotube via the vacant defect when the size of the defect is more than the threshold. The vacant defect size control allows the researchers to extract molecular hydrogen from a mixture of gas and stock it up inside the carbon nanotubes. The physisorption was found to be favored by high pressure and low temperature. Moreover, the storage proficiency of the materials was also enhanced by the introduction of more defects, which in practice is the reduction in carbon number of single-walled carbon nanotubes [5].

In an investigation, a combination study with semi-empirical research, force-matching and theoretical method was made for the determination of the optimal hydrogen adsorption ability of an open-ended single-walled carbon nanotube, as a function of diameter. Semi-empirical analysis was done to establish the adsorption isosteric heat and monolayer coverage value from existing thermodynamic statistics. Quantum mechanical studies were performed for the determination of adsorption energy on the inner and outer surfaces of the single-walled carbon nanotubes. Lastly, force-matching and classical Lennard–Jones potential standard was used for determining the values of specific surface area, monolayer coverage, and adsorption capacity. The physisorption of hydrogen was analyzed on zigzag single-walled carbon nanotubes at temperature of −40 °C and 25 °C and pressures of 0.1 to 10 MPa. All the data of the adsorption were explained via Toth model. The exterior physisorption energy of single-walled carbon nanotubes was in the range of 1.35–1.62 kcal/mol, whereas the interior physisorption energy ranges from 1.22 to 2.43 kcal/mol at optimized diameter of 8–12 Å. At the optimized conditions and diameter, 1.75 wt. % of hydrogen adsorption was obtained [20].

7.4 Physisorption of Hydrogen and Graphene

Graphene and graphene-based nanomaterials are actively involved in hydrogen adsorption studies. It is understood that the large surface area; defect sites such as carbon vacancy, wrinkles, and sheet edges; and presence of π-electron density out-of-plane of graphene sheet favor physisorption of incoming gas molecules. Therefore, the hydrogen uptake capacity of graphene depends on how it has been processed. Some of the studies that involve use of graphene for adsorption of hydrogen are given in Table 7.1.

7.5 Physisorption of Hydrogen and Metallic Nanomaterials

Different metallic nanomaterials either in pure form or in the form of nanocomposites are proving themselves useful in the area of hydrogen storage. In this section, different metallic nanomaterials involved in the physisorption of hydrogen are discussed.

Metal organoclays are also used in hydrogen storage. A study has reported the metal organoclays fabricated with integration of Boltorn polyol dendrimer H_{30} (a polyalcohol) in Na^+-exchanged montmorillonite (NaMt). The polyalcohol addition in the substrate clay was followed by in situ loading of Cu (0) and Pd (0) nanoparticles. The polyalcohol-modified organoclays showed promising ability of CO_2 retention (3.6–11.1 μmolg^{-1}) because of several OH^- groups of the added alcohol. However, the integration of the metals in the clay drastically reduced the CO_2 retention but caused a momentous increase in hydrogen uptake (51.8–508.2 μmolg^{-1}) where OH^- groups served as the metal-stabilizing sites. The metal stabilization occurs because of the increase in the formation of HO: Cu (0) and HO: Pd (0) associations and decrease in HO: CO_2 interactions. The hydrogen and carbon dioxide retaining abilities of the prepared nanomaterials were found to be closely associated with number of hydroxyl groups of the incorporated polyalcohol moiety. The interactions of hydrogen were

Table 7.1 Graphene and its derivative involve in storage of hydrogen

Graphene material	Description	% Adsorption of H_2	References
Graphene and carbon nanotube composites	Physisorption of H_2 on graphene surface and on exterior of carbon nanotubes	8.4 wt.%	[20]
Li-adsorbed graphene	Li ions turn semi-conductive graphene to good metal substance	12.8 wt. %	[21]
Ti on graphene oxide	Ti atoms were able to bind with hydrogen molecules	4.9 wt. %	[22]
Ti graphene, Li graphene	Strain on graphene caused increased adsorption of metals, hence hydrogen	9.5 wt. %, 15.4 wt.%, respectively	[23]
Pd-decorated graphene, Pt-decorated graphene	Pt and Pd nanoparticles fill the voids in graphene	0.15 wt.% and 0.156 wt.%, respectively	[24]
Ni–B nanoalloy-doped graphene	Graphene was doped with nanoalloy by chemical reduction method	4.4 wt.%	[25]
Ca-restrained-B-doped graphene	Clustering issue of Ca-doped graphene was resolved by doping the nanomaterials with boron	8.38 wt.%	[26]

found to be of physical nature as an easy release of gas was found at 20 °C and 75 °C or even at room temperature under vacuum. This demonstrates unequivocally the reversible capture of hydrogen. Room temperature gas release was also obtained under vacuum. There was an increase in hydrogen uptake with the increase in contact time. This is the evidence of diffusion due to structure compaction caused by metal integration. The study proposed the prepared nanomaterials as low cost, biodegradable, and reversible hydrogen storage materials with remarkable future prospects [27].

Zhao et al. have used three activated carbon units with visible surface areas of 2450 to 3200 m^2g^{-1} doped with varying concentrations of Pd nanoparticles for the hydrogen adsorption applications. The activated carbons were integrated with varying concentrations of Pd nanoparticles ranging from 1.3 to 10.0 wt.%. The hydrogen adsorption capabilities of the nanomaterials were determined at 8 MPa pressure and −193 °C and 25 °C temperatures. The study found that the hydrogen loading is dependent upon concentration of Pd nanoparticles and high pressure of 2–3 MPa at room temperature. The storage ability of the material is less than <0.2 wt.% below these conditions. The hydrogen storage capacity is controlled by volume of micropores at higher pressure. Pd nanoparticles doping at −193 °C exhibited negative effect on hydrogen adsorption regardless of the pressure applied. The prepared nanomaterials were characterized with nitrogen adsorption at −193 °C, temperature-programmed reduction (TPR), X-Ray diffraction (XRD), transmission electron microscopy (TEM), and hydrogen chemisorption with higher surface area which produced smaller metallic nanoparticles at a given Pd content [28].

In an investigation, computational studies have proposed enhanced hydrogen adsorption abilities of carbon–metal nanocomposites. The study determined the hydrogen adsorption capacity of arranged porous carbon material supporting Pd nanoclusters. The said nanomaterials were attained via chemical impregnation of carbon template with the solution of H_2PdCl_4 followed by subsequent reduction. In the carbon porous material, 10 wt.% of palladium nanoclusters with average size of 2 nm was homogeneously dispersed. Hydrogen isotherm adsorption calculations and thermal desorption spectroscopy (TDS) characterization were made for the determination of thermodynamic hydrogenation properties of pristine carbon template as well as the 10 wt.% Pd nanocluster loaded carbon templates. The carbon templates with Pd nanoparticles do not increase the hydrogen uptake at −193 °C temperature and 1.6 MPa pressure. In these conditions, the hydrogen was being stored via physisorption; however, at 25 °C and 0.5 MPa moderate pressure the carbon templates of 10 wt.% nanocluster of Pd depicted eight times higher uploading of the hydrogen as compared to the pristine carbon support. TDS analysis credited rapid increase of hydrogen adsorption to the Pd nanoclusters [29].

The hydrogen adsorptions on metallo-carbohedrene Ti_8C_{12} and nanocrystals of $Ti_{14}C1_3$ were investigated using first-principles calculations. The formation of carbon hydrides was not possible in the absence of Ti atoms which acts as the catalyst to facilitate the dissociation of hydrogen. Titanium atoms present on nanocarbide surface form complex coordinate with several molecular hydrogen ligands. Hydrogen adsorption capabilities of 6.1 wt.% 7.7 wt. % for Ti_8C1_2 and for $Ti_{14}C1_3$, respectively, were achieved with above than 80% of the hydrogen adsorbed in the energy ranging 0.17–0.89 eV per molecule of hydrogen. After the formation of macroscopic substance by the nanoparticles (due to hydrogen adsorption), chemisorption of the hydrogen starts declining; however, more hydrogen molecules are adsorbed via physisorption. The study suggested TiC nanoparticles as potential hydrogen storage material at approximate ambient conditions [30].

7.6 Conclusion

Physisorption is a remarkable hydrogen storage phenomenon. Physisorption allows the adsorption of the hydrogen on various material surfaces via van der Waals linkages. Nanomaterials have revolutionized the area of hydrogen storage via physisorption because of improved storage capacity. Different nanomaterials like nanotubes and fullerenes allow the adsorption of hydrogen not only on their surface but also inside the nanomaterials. The advanced research is needed for the utilization of these nanomaterials in portable applications of hydrogen storage.

References

1. Han SW, Cha G-B, Park Y, Hong SC (2017) Hydrogen physisorption based on the dissociative hydrogen chemisorption at the sulphur vacancy of MoS_2 surface. Sci Rep 7(1):7152. https://doi.org/10.1038/s41598-017-07178-9
2. Basile A, Iulianelli A (2014) Advances in hydrogen production, storage and distribution. Elsevier
3. Klontzas E, Tylianakis E, Froudakis GE (2008) Hydrogen storage in 3D covalent organic frameworks. A multiscale theoretical investigation. The Journal of Physical Chemistry C 112 (24):9095–9098
4. Pupysheva OV, Farajian AA, Yakobson BI (2008) Fullerene Nanocage Capacity for Hydrogen Storage. Nano Lett 8(3):767–774. https://doi.org/10.1021/nl071436g
5. Sun G, Tangpanitanon J, Shen H, Wen B, Xue J, Wang E, Xu L (2014) Physisorption of molecular hydrogen on carbon nanotube with vacant defects. J Chem Phys 140(20):204712
6. Kajiura H, Tsutsui S, Kadono K, Kakuta M, Ata M, Murakami Y (2003) Hydrogen storage capacity of commercially available carbon materials at room temperature. Appl Phys Lett 82(7):1105–1107
7. Chahine R, Bose TK (1994) Low-pressure adsorption storage of hydrogen. Int J Hydrogen Energy 19(2):161–164
8. Lee J, Farha OK, Roberts J, Scheidt KA, Nguyen ST, Hupp JT (2009) Metal–organic framework materials as catalysts. Chem Soc Rev 38(5):1450–1459
9. Férey G, Latroche M, Serre C, Millange F, Loiseau T, Percheron-Guégan A (2003) Hydrogen adsorption in the nanoporous metal-benzenedicarboxylate M(OH)(O2C–C6H4–CO2) (M = Al3+, Cr3+), MIL-53. Chem Commun 24:2976–2977. https://doi.org/10.1039/B308903G
10. Isaacson CW, Kleber M, Field JA (2009) Quantitative analysis of fullerene nanomaterials in environmental systems: a critical review. Environ Sci Technol 43(17):6463–6474. https://doi.org/10.1021/es900692e
11. Wang Z, Yao M, Pan S, Jin M, Liu B, Zhang H (2007) A barrierless process from physisorption to chemisorption of H2 molecules on light-element-doped fullerenes. J Phys Chem C 111(11):4473–4476
12. Mert H, Deniz CU, Baykasoglu C (2020) Monte Carlo simulations of hydrogen adsorption in fullerene pillared graphene nanocomposites. Mol Simul 46(8):650–659. https://doi.org/10.1080/08927022.2020.1758696
13. Sathe RY, Bae H, Lee H, Dhilip Kumar TJ (2020) Hydrogen storage capacity of low-lying isomer of C24 functionalized with Ti. Int J Hydrogen Energy 45(16):9936–9945. https://doi.org/10.1016/j.ijhydene.2020.02.016
14. Ozturk Z, Baykasoglu C, Kirca M (2016) Sandwiched graphene-fullerene composite: A novel 3-D nanostructured material for hydrogen storage. Int J Hydrogen Energy 41(15):6403–6411. https://doi.org/10.1016/j.ijhydene.2016.03.042
15. Rather Su (2020) Preparation, characterization and hydrogen storage studies of carbon nanotubes and their composites: a review. Int J Hydrogen Energy 45(7):4653–4672. https://doi.org/10.1016/j.ijhydene.2019.12.055
16. Cabria I, López MJ, Alonso JA (2006) Density functional calculations of hydrogen adsorption on boron nanotubes and boron sheets. Nanotechnology 17(3):778
17. Dillon AC, Jones KM, Bekkedahl TA, Kiang CH, Bethune DS, Heben MJ (1997) Storage of hydrogen in single-walled carbon nanotubes. Nature 386(6623):377–379
18. Ye Y, Ahn CC, Witham C, Fultz B, Liu J, Rinzler AG, Colbert D, Smith KA, Smalley RE (1999) Hydrogen adsorption and cohesive energy of single-walled carbon nanotubes. Appl Phys Lett 74(16):2307–2309
19. Chambers A, Park C, Baker RTK, Rodriguez NM (1998) Hydrogen storage in graphite nanofibers. J Phys Chem B 102(22):4253–4256
20. Kosasih EA, Kurniawan B, Zulkarnain IA (2016) Optimization of hydrogen storage capacity by physical adsorption on open-ended single-walled carbon nanotube as diameter function. Int J Technol 7(2):264–273

21. Ataca C, Aktürk E, Ciraci S, Ustunel H (2008) High-capacity hydrogen storage by metallized graphene. Appl Phys Lett 93(4):043123. https://doi.org/10.1063/1.2963976
22. Wang L, Lee K, Sun YY, Lucking M, Chen Z, Zhao JJ, Zhang SB (2009) Graphene oxide as an ideal substrate for hydrogen storage. ACS Nano 3:2995–3000
23. Zhou M, Lu Y, Zhang C, Feng YP (2010) Strain effects on hydrogen storage capability of metal-decorated graphene: a first-principles study. Appl Phys Lett 97(10):103109
24. Huang C-C, Pu N-W, Wang C-A, Huang J-C, Sung Y, Ger M-D (2011) Hydrogen storage in graphene decorated with Pd and Pt nano-particles using an electroless deposition technique. Sep Purif Technol 82:210–215. https://doi.org/10.1016/j.seppur.2011.09.020
25. Wang Y, Guo CX, Wang X, Guan C, Yang H, Wang K, Li CM (2011) Hydrogen storage in a Ni–B nanoalloy-doped three-dimensional graphene material. Energy Environ Sci 4(1):195–200
26. Beheshti E, Nojeh A, Servati P (2011) A first-principles study of calcium-decorated, boron-doped graphene for high capacity hydrogen storage. Carbon 49(5):1561–1567. https://doi.org/10.1016/j.carbon.2010.12.023
27. Bouazizi N, Barrimo D, Nousir S, Ben Slama R, Shiao TC, Roy R, Azzouz A (2018) Metal-loaded polyol-montmorillonite with improved affinity towards hydrogen. J Energy Inst 91(1):110–119. https://doi.org/10.1016/j.joei.2016.10.002
28. Zhao W, Fierro V, Zlotea C, Izquierdo MT, Chevalier-César C, Latroche M, Celzard A (2012) Activated carbons doped with Pd nanoparticles for hydrogen storage. Int J Hydrogen Energy 37(6):5072–5080. https://doi.org/10.1016/j.ijhydene.2011.12.058
29. Campesi R, Cuevas F, Gadiou R, Leroy E, Hirscher M, Vix-Guterl C, Latroche M (2008) Hydrogen storage properties of Pd nanoparticle/carbon template composites. Carbon 46(2):206–214. https://doi.org/10.1016/j.carbon.2007.11.006
30. Zhao Y, Dillon AC, Kim Y-H, Heben MJ, Zhang SB (2006) Self-catalyzed hydrogenation and dihydrogen adsorption on titanium carbide nanoparticles. Chem Phys Lett 425(4):273–277. https://doi.org/10.1016/j.cplett.2006.05.034

Chapter 8
Chemisorption

8.1 Introduction

As mentioned in the previous chapter, one of the major reasons of hydrogen not be used as major fuel is due to the limitations in storage of hydrogen. The storage of the hydrogen ought to be in way that it should be reversible at ambient conditions, or it can be discharged from the storage medium as required. Chemical adsorption or chemisorption is another mechanism of hydrogen storage. In terms of hydrogen storage, any material when gets attached to the hydrogen becomes a hydride. Over the last few years, the scope of materials to get hydrogenated or hydride formation has immensely expanded. Simple hydrides subsequently shifted toward complex hydrides leading to the formation of chemical hydrides. Hydrides are now expanding from left to the right of the periodic table [1].

Some elements make ionic bonds with hydrogen, many of them combines with metallic bonds whereas rest of them combine through covalent bonding. The major chemical substances and compounds that are employed in generation of hydrogen storage are metal hydrides [2], alanes, and alanates [3]; complex hydrides, borohydrides [4]; and imides and amides [5]. In contrast to physisorption, the chemisorption is capable of storing greater amounts of hydrogen per unit of volume and mass. The major drawback associated to this mode of storage is less reversibility opportunities. Chemisorption most of the times requires elevated temperatures to discharge the absorbed gas with lower kinetics.

Despite the restraints, it is now agreed after several years of investigations and research in this area that the most favorable hydrogen storage pathway is the usage of solid substances that can physically adsorb or can chemically react with hydrogen at volume densities larger than that of liquid hydrogen. Three contrasting characteristics must be met at same time of the material to be considered as the potential hydrogen storage medium. First, they should have high volumetric and gravimetric hydrogen loading densities, secondly the complete reversibility of the stored gas should be attained at ambient conditions or minimum at moderate temperature. Lastly, the

S. Farrukh et al., *Nanotechnology and the Generation of Sustainable Hydrogen*, Green Energy and Technology, https://doi.org/10.1007/978-3-030-60402-8_8

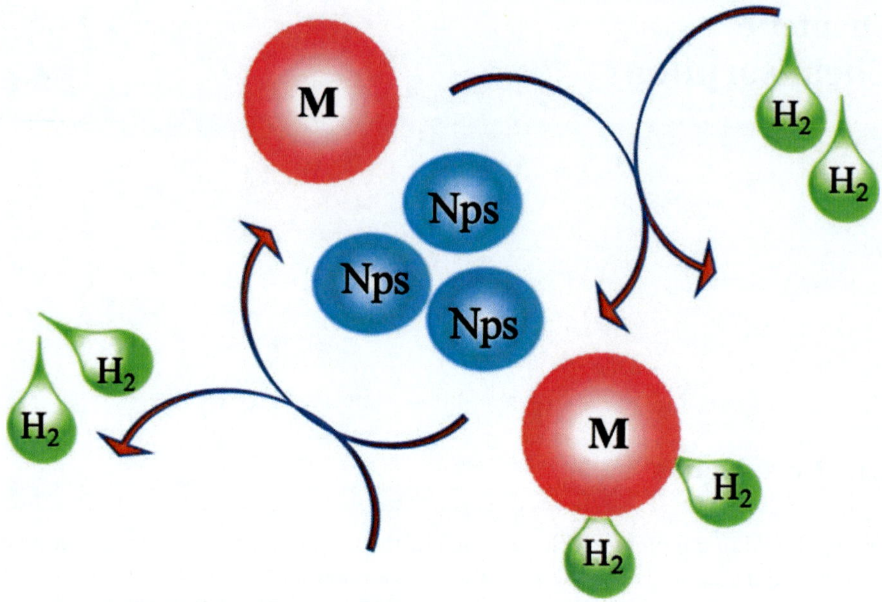

Fig. 8.1 Simple mechanism of chemical hydrogen adsorption and desorption with nanomaterials (where Nps = nanoparticles, M = hydride-forming metal)

material should have high charge per release rates. Additionally, there are some other important requirements that should also be fulfilled for the material to be used as hydrogen storage medium, like stability, thermal conductivity, suitable cost, durability, and no poisoning [6]. A general mechanism of chemical hydrogen adsorption and desorption is given in Fig. 8.1.

The advancement in the formation of hydrides is accompanied by the fast development in the nanotechnology. The coupling of the both has led to the remarkable progress in the field of hydrogen storage. This chapter discusses important nanomaterial with the ability to form hydrides as hydrogen storage.

8.2 Chemisorption of Hydrogen and Fullerenes

Fullerenes are nanomaterials with sp^2 hybridized hollow carbon cages. Commonly found as C_{60}, but other order materials like C_{70}, C_{78}, C_{28}, C_{36} are also known [7]. Fullerenes are actively involved in the chemisorption of hydrogen. Different studies have been made and reported the use of fullerenes in chemisorption of hydrogen.

In an investigation, the researchers used first-principles electronic structure calculations for depicting the trapping-mediated dissociative chemisorption of hydrogen. The chemisorption of the hydrogen was allowed to take place on boron-doped fullerenes. They are formed by the replacement of six atoms of carbon from C_{60} with six boron atoms making it $C_{54}B_6$. The doped nanomaterial had D_{3d} symmetry. By employing van't Hoff–Arrhenius law and Polanyi–Wigner equation, it was optimized that that the hydrogen will get adsorbed on boron-doped fullerenes and can undergo dissociation without additional catalysts. The separation of the hydrogen takes place at ~0.5 picoseconds at 25 °C. The study reported the boron-doped fullerenes as a good storage material for atomic hydrogen and poor for molecular hydrogen [8]. A seemingly similar study reported the use of B and Be-doped fullerenes for hydrogen adsorption [9]. The fabrication of boron-doped fullerene is reported by *Ting el al.* They can be prepared by the laser vaporization treatment of a boron nitride and graphite mixture. The mixture is used in the form of composite disk of both materials bound with cement [10].

ZhiGang et al. have reported the barrierless physisorption to chemisorption of hydrogen molecules on the fullerenes doped with lightweight elements. The study employed local density approximation (LDA) within density functional theory (DFT) as process. The DFT process is capable of large systems at low computational costs. In practice, $C_{35}B$ fullerene is brought onto the space of $C_{35}B-H_2$ stable system. The two $C_{35}B$ molecules now behave like forceps with each attaching itself to the opposite side of the small H_2 molecules resulting in its dissociation to H, H (with one H with each of $C_{35}B$). This is barrierless physisorption to chemisorption. The study suggested that H–H bond breaking occurs because of energy discharge as $C_{35}B$ gets near to the $C_{35}BH_2$. The evolved energy increased the bond length between H–H ultimately leading to chemisorption of H_2 molecules in the hydrogen storage method. Overall the energy decreases monotonously during this procedure [11].

Ni-dispersed fullerenes exhibited that they can be a promising alternative to the reversible hydrogen storage. In a study, it is exhibited that fullerenes covered with nickel on surface are capable of storing three molecules of hydrogen. Therefore, fullerenes with dense coating of Ni are thought of excellent storage medium for hydrogen storage, with the ability to store ~6.8 wt.% of hydrogen. 6.5 wt. % of storage is the target that Department of Energy is trying to achieve for automotive purposes. The study reported that the Ni dispersed has 11.8 kcal/mol of hydrogen desorption activation barrier, which is suitable for several practical H_2 storage purposes [12].

Charged fullerenes are also high-capacity hydrogen storage medium. In an investigation, the DFT was used for making first-principle calculations and the potential of charged fullerenes for hydrogen storage was analyzed. It was found that both positively and negatively charged fullerenes depict remarkable increase (0.18−0.32 eV) in H_2 storage capacity at 25 °C. The improved binding of the hydrogen is delocalized and covered the whole surface of fullerene. When completely covered with hydrogen the charged fullerene has ~ 8 wt.% hydrogen coverage. The study showed that the enhancement in hydrogen storage of fullerenes due to metals is less as compared to the charged fullerene because metals usually get trapped inside the fullerene cage, and there is less charge distribution on outer surface [13].

A study conducted by *HongJiang* et al. has reported alkali metal-doped fullerene for the chemisorption of hydrogen. Fullerene doped with Li (Li_6C_{60}) and Na (Na_6C_{60}) proved as promising hydrogen storage materials. However, the stabilities, structures, and electronic characteristics are still ambiguous. *HongJiang* et al. have used four isomers of Na_6C_{60} and Li_6C_{60} and analyzed them for their hydrogen storage capacity with DFT model. The effect of pressure and temperature was determined in all the isomers by employing the statistical thermodynamic methods. For the determination of hydrogen storage capacity, the doped fullerenes were allowed to interact with 36 atoms of hydrogen at 300 ^{0}C and 100 bar pressure. The hydrogen binding on the surface of fullerenes followed two mechanisms with different binding energies. In the first mechanism, the hydrogen atoms are adsorbed directly onto the surface of C_{60} carbon atoms resulting in the formation of chemical bond. The other mechanism involves the hydrogen molecules which gets physically adsorbed onto lithium or sodium atoms. Hydrogen binds in the molecular form via charge polarization mechanism as the Li and Na are positively charged because of their metallic nature. The study concluded that Li-doped fullerene has better hydrogen storage capacity compared to the Na-doped fullerene because of greater stability of the Li-doped fullerene. The highest gravimetric density shown by one of the Li-doped fullerene is 4.5 wt.% where it is 4 wt. % for sodium-doped material [14].

8.3 Chemisorption of Hydrogen and Carbon Nanotubes

Carbon nanotubes are involved actively in the storage of hydrogen. Several studies have reported the use of carbon nanotubes in chemisorption of hydrogen. *Nikitin* et al. have investigated chemisorption of hydrogen in single-walled carbon nanotubes. In order to determine whether these nanotubes can be used for hydrogen chemisorption, atomic hydrogen was used for the hydrogenation of the nanotubes. The study concluded that the maximum hydrogen adsorption of the nanotubes is dependent on diameter of nanotube. The study claimed that the nanotubes with the diameter of ~2.0 nm can have up to 100% hydrogenation, with complete stability at 25 °C. This implies that the particular carbon nanotubes can have 7 wt. % hydrogen storage capacity via the establishment of reversible C-H bonds [15].

A study has reported the dissociative chemisorption of hydrogen on carbon nanotubes depending upon the first-principle calculations. The chemisorption of hydrogen is the result of C-H bond formation after the breakage of the H–H bond, when high pressure is applied on the two neighboring carbon nanotubes. The chemisorbed hydrogen is released when the applied pressure is removed from the nanotubes, hence making the process reversible. Fullerenes also depicts similar type of behavior in vicinity of hydrogen when pressure is applied [16].

There are three possible types of geometries shown by carbon nanotubes after the adsorption of hydrogen. The first is arch-type geometry which is shown by the nanotubes when the ions of hydrogen get adsorbed to the top spots of carbon atoms. This chemisorption shown by hydrogen ions is exothermic process. The hydrogen gets adsorbed on the exterior of the carbon nanotubes where each hydrogen atom gets attached to the carbon atom with sp^3 hybridization. This arrangement of the atoms increases the diameter of the carbon nanotubes to 6.88 nm–7.78 nm. Another type of geometry is zigzag, where adsorption of hydrogen takes place both inside and outside of the carbon nanotubes giving the structure a zigzag appearance. Owing to less strain on the C–C bonding this geometry is regarded as more stable than the arch type. There is comparatively less increase in the average diameter to 7.13 nm. This type of geometry allows large amounts of the hydrogen storage which can increase with the increase in diameter [17]. However, the greater uptake of hydrogen can produce repulsive energies resulting in the breakage of the nanotube walls. *Lee* et al. reported that repulsive energies determine the stability and hydrogen uptake ability of the carbon nanotubes [18].

Carbon nanotubes of varying purity were analyzed in a study, for their ability of hydrogen adsorption at 25 MPa pressure and -196 °C temperature. It was found that the carbon nanotubes with high purity can adsorb greater amount of hydrogen as compared to the less pure carbon nanotubes. Additionally, in case of multiwalled carbon nanotube, the enhanced adsorption of the hydrogen in highly pure multiwalled carbon nanotubes is associated to high surface area and more available adsorption spots [19].

In a computational investigation, first-principle calculations were made within density functional theory (DFT) for the determination of hydrogen chemisorption on carbon nanotubes.

The study elaborated chemisorption of one and two hydrogen atoms on the exterior walls of varying armchair single-walled carbon nanotubes. The results of the study showed that the two hydrogen atoms prefer bond formation at adjacent sites instead of alternate carbon positions [21]. Different results on zigzag nanotubes are reported by *Dinadayalane* et al. [22] Commonly, the exothermicity of chemisorption of hydrogen decreases with the increase in the diameter of armchair nanotubes, whereas the opposite is true for hydrogen chemisorption on zigzag nanostructures. The adsorption of one and two hydrogen atoms expressively change the C-C bond distance of the carbon nanotube. In the presence of hydrogen atoms, the carbon atom(s) adsorbs the hydrogen atom(s) resulting in the alteration of hybridization of carbon from sp^2 to sp^3 at the chemisorption spot(s) [21].

Different studies have reported modified carbon nanotubes for hydrogen storage applications. For instance, the metal-doped carbon nanotubes have depicted promising hydrogen adsorption. In some cases, the ability of hydrogen storage shown 40 times more hydrogen storage ability as compared to the pure carbon nanotubes. The improved hydrogen adsorption can be associated to the fact that in modified carbon nanotubes the initial hydrogen adsorption is done by the metallic nanoparticles, resulting in the subsequent dissociation of the hydrogen molecule. The atoms thus formed are then and spilled to the nanotubes. A precise information of some of metal-modified nanotubes involved in hydrogen storage is given in Table 8.1.

Hydrogen adsorption ability at ambient pressure and room temperature in varying types of carbon nanotubes is very less than the target aimed by the USA Department of Energy. However, it can be inferred form the data in Table 8.1 that at low temperature such as -196 °C and high pressure like 4 MPa, some carbon nanotubes offer hydrogen storage equal or above the target of energy department. Still that carbon nanotubes it

Table 8.1 Summary of reported hydrogen storage capacity in activated, nanostructured metaland metal oxide and hydride CNT composites. (Adapted with permission from Ref [20], Copyright, Elsevier 2020)

Sample name	Hydrogen storage (wt.%)	Temperature (K)	Pressure (MPa)
KOH-acti-MWCNTs	0.32	298	1.88
KOH-acti-MWCNTs	4.47	298	0.1
Acti-MWCNTs	1.0	293	10
NaOH-acti-CNTs	4.5	77	4
Pd-MWCNTs	0.18	298	1.6
TiO_2-MWCNTs	0.4	298	1.8
Ag-MWCNTs	0.86	298	2.3
Pt–Pd alloy	−2.0	298	2.0
Pt-MWCNTs	2.9	298	1.67
Ni-MWCNTs	2.27	298	8.0
MgH_2-MWCNTs	5.9	673	4.6
Ti-MWCNTs	2.0	298	1.6
Co-Oxide/MWCNTs	−0.8	298	2.3
Cu-Oxide/MWCNTs	−0.9	298	2.3
Mn-Oxide/MWCNTs	−0.94	298	4.0
Ti-Mn-Cr/CNTs	4.6	353	–
$LiBH_4$. NH_3/CNTs	6.7	553	–
Pd-CNTs	0.66	298	2.0
V-CNTs	0.69	298	2.0
Fe-Ag/TiO_2/MWCNTs	10.94	298	–

is difficult to use carbon nanotubes for commercial applications because of obvious reasons [20].

8.4 Physisorption of Hydrogen and Graphene

Graphene and graphene-based nanomaterials are actively involved in hydrogen adsorption studies. It is understood that the large surface area, defect sites such as carbon vacancy, wrinkles, sheet edges, and presence of π-electron density out-of-plane of graphene sheet favor physisorption of incoming gas molecules. Therefore, the hydrogen uptake capacity of graphene depends on how it has been processed. Some of the studies that involve use of graphene for adsorption of hydrogen are given in Table 8.2.

Table 8.2 Graphene-based nanomaterials for adsorption of hydrogen

Graphene material	Description	Reference
Ca on porous 3D graphene	Semiconductor nature of the graphene is changed due to the gain of electrons from calcium. 8.4 wt.% adsorption of hydrogen was achieved	[23]
Chemically altered graphene with changed configuration of local binding	Molecular simulations were used to analyze basic effects of hydrogen chemisorption on mechanical properties of graphene	[24]
Boron-doped graphene	The chemisorption method of H_2 adatoms on B-doped graphene was analyzed via ab initio calculations	[25]
Ca on graphene	Kubas bonding takes place between Ca and H_2. 5–6 wt.% adsorption of hydrogen was achieved	[26]
Graphene-supported Al clusters	Mono-vacant defective graphene was chemically bonded with Al clusters	[27]
Graphene-N_2 decorated with Pd clusters	Buckling induced by the chemisorption of hydrogen, resulting in the diminishing of dissociation of H_2	[28]

8.5 Chemisorption of Hydrogen and Metallic Nanomaterials

Several metallic nanomaterials are actively involved in the chemisorption of hydrogen. From the data analysis in Table 8.1, it is evident that several light metals as well as transition metals are involved in nanocomposite formation along with the carbon nanotubes. This highlights the importance of metallic nanomaterials in the hydrogen storage. Here metallic nanomaterials with hydrogen storage applications, other than those that are used with the carbon nanotubes, will be discussed.

Metal hydrides are promising option for the storage of hydrogen. It can be stored effectively in the form of chemical compounds as metal hydrides. The hydrogen atoms dissociate during metal hydride formation and are inserted in spaces inside the metal lattice. The metals are capable of reacting directly with the hydrogen at a particular condition of temperature and pressure. The chemical reaction can be given as [6].

$$M + H_2 \rightarrow MH_2 + \Delta H$$

The formation of the hydride is an exothermic and reversible process. The adsorbed hydrogen in these compounds can be released easily upon increasing the temperature or/and lowering the pressure. There are several advantages of hydrides as hydrogen storage medium such as they can surpass the limitations like large volume, low temperature, high pressure, and high evaporation losses. However, it is of utmost importance to use the appropriate metal for the purpose [29]. Mostly the hydrides with high gravimetric and volumetric capacities have been investigated and employed as the possible hydrogen medium. Almost all of the hydrides have higher volumetric capacities. However, magnesium hydrides are only one with required gravimetric storage abilities, hence making them a promising option for the application.

In an investigation, researchers employed nanostructured magnesium hydride along with a $ZrFe_{1.4}Cr_{0.6}$ catalyst for hydrogen storage purposes. The study claimed that the hydride phases exhibit the remarkable desorption kinetics. In 15 min at 280 °C, 80% of the maximum capacity was desorbed. The proposed mechanism for the desorption of hydrogen from the nanohydrides suggests the first half of the reaction to be nucleation orientated, whereas the major part of the reaction is growth controlled via the hydrogen atom diffusion [30].

MOFs are nanoporous substances with high specific area and low density. They are formed by the linking of metal oxide species with the organic moieties [31]. Metal organic frameworks (MOFs) are also involved in the chemisorption of hydrogen. In a study, considerable hydrogen storage was achieved with MOF-5 and IRMOF-8 via hydrogen splitting and spillover [32]. Another study has also reported Pd nanoparticles integrated in MOFs for hydrogen adsorption properties [33].

Chenggang et al. have studied hydrogen successive dissociative chemisorption on sub-nanopalladium clusters. The study was made using first-principle calculations within DFT. The lowest energy clusters from Pdn (where n = 2–9) were selected for the study. The chemisorption of dissociative hydrogen and the following migration of hydrogen atoms on the clear Pd clusters were determined to be barrierless. The energy chemisorption dissociation and desorption of hydrogen decrease with the increase in the spread of hydrogen atoms, hence resulting in the decrease of catalytic efficiency of nanomaterials. At complete saturation of clusters, these energy changes were identified to vary in small ranges irrespective of the size of cluster size. As loading of hydrogen on the sub-nanoclusters increases, the nature of bonding in the clusters slowly shifts from metallic to covalent. The study also determined that with the increase in size of the clusters the adsorption of hydrogen increases. Moreover, it was determined that the ability of Pd clusters for hydrogen loading is smaller as compared to the Pt clusters [34].

Nanoalloys have also been proven useful for hydrogen storage purposes. Different nanoalloys particularly of Pd are reported in literature for their use as hydrogen storage material. *Abdelmalek* et al. have reported the use of Pd-Ir nanoalloys restrained on mesoporous carbon for the hydrogen storage purposes. The average size of nanoalloys depends on the composition and the metal in the alloys. In the material under consideration, the size of the particle is in the range of 2.7–3.5 nm. The structural analysis suggests that the structure of the nanomaterials is face-centered cubic. The hydrogen storage properties of the nanoalloy can be controlled by adjusting the chemical composition of the constituent metals. The nanoalloys with greater amount of Pd nanoparticles exhibited hydrogen adsorption behavior, whereas the nanoalloys with greater amount of Ir nanoparticles do not depict hydrogen adsorption under ambient conditions of pressure and temperature. The hydride formed with the nanomaterials depicted difference in thermodynamic properties to the bulk counterparts. Additionally, the nanoalloys (with more Pd nanoparticles) depicted greater hydrogen absorption capability as compared to the similar composition bulk alloy. This can be attributed to the greater surface area of nanoscale materials [35]. Several other nanoalloys like Pd–Co nanoalloy supported on mesoporous carbon [36], Pd–Ni nanoparticles dispersed on carbon template, Pd–Au and Pd–Rh nanoparticles supported on polyvinyl pyrrolidone polymer, Pd–Hg nanoalloys supported on carbon foam, and Pd–Cd and Pd–Ag nanoalloys immobilized on activated carbon are also reported for their hydrogen adsorption abilities [37–41, 29].

8.6 Conclusion

Chemisorption is another promising means of hydrogen storage. Chemisorption allows the development of strong chemical linkages between the hydrogen and the adsorbent. In the past years, the chemisorption of hydrogen has been improved by utilizing the advanced hydrides in combination with the nanomaterials. The carbon-based nanomaterials are the most widely used for the chemisorption of hydrogen. The current study suggests that the researches should be conducted for the development of economical routes for preparation of ultrapure (purity enhances the adsorption of hydrogen) carbon-based nanomaterials. The required nanomaterials are very expensive, thus making them impossible to use for practical applications.

References

1. Basile A, Iulianelli A (2014) Advances in hydrogen production, storage and distribution. Elsevier
2. Agro S, DeCarmine T, DeFelice S, Thoma L (2005) Annual Progress Report for the DOE Hydrogen Program, 2005. US Department of Energy (DOE) website: https://www.hydrogen energy gov:790
3. Sandrock G, Reilly J, Graetz J, Zhou WM, Johnson J, Wegrzyn J (2005) Accelerated thermal decomposition of AlH 3 for hydrogen-fueled vehicles. Appl Phys a 80(4):687–690
4. Vajo JJ, Skeith SL, Mertens F (2005) Reversible storage of hydrogen in destabilized LiBH4. J Phys Chem B 109(9):3719–3722
5. Liu Y, Zhong K, Gao M, Wang J, Pan H, Wang Q (2008) Hydrogen storage in a LiNH2–MgH2 (1: 1) system. Chem Mater 20(10):3521–3527
6. Satyapal S, Petrovic J, Read C, Thomas G, Ordaz G (2007) The U.S. Department of Energy's National Hydrogen Storage Project: Progress towards meeting hydrogen-powered vehicle requirements. Catal Today 120 (3):246–256. https://doi.org/10.1016/j.cattod.2006.09.022
7. Isaacson CW, Kleber M, Field JA (2009) Quantitative analysis of fullerene nanomaterials in environmental systems: a critical review. Environ Sci Technol 43(17):6463–6474. https://doi.org/10.1021/es900692e
8. Lee H, Li J, Zhou G, Duan W, Kim G, Ihm J (2008) Room-temperature dissociative hydrogen chemisorption on boron-doped fullerenes. Phys Rev B 77(23):235101

9. Kim Y-H, Zhao Y, Williamson A, Heben MJ, Zhang SB (2006) Nondissociative adsorption of H 2 molecules in light-element-doped fullerenes. Phys Rev Lett 96(1):016102
10. Guo T, Jin C, Smalley RE (1991) Doping bucky: formation and properties of boron-doped buckminsterfullerene. J Phys Chem 95(13):4948–4950
11. Wang Z, Yao M, Pan S, Jin M, Liu B, Zhang H (2007) A barrierless process from physisorption to chemisorption of H2 molecules on light-element-doped fullerenes. J Phys Chem C 111(11):4473–4476
12. Shin WH, Yang SH, Goddard Iii WA, Kang JK (2006) Ni-dispersed fullerenes: hydrogen storage and desorption properties. Appl Phys Lett 88(5):053111
13. Yoon M, Yang S, Wang E, Zhang Z (2007) Charged fullerenes as high-capacity hydrogen storage media. Nano Lett 7(9):2578–2583. https://doi.org/10.1021/nl070809a
14. Ren H, Cui C, Li X, Liu Y (2017) A DFT study of the hydrogen storage potentials and properties of Na-and Li-doped fullerenes. Int J Hydrogen Energy 42(1):312–321
15. Nikitin A, Li X, Zhang Z, Ogasawara H, Dai H, Nilsson A (2008) Hydrogen Storage in Carbon Nanotubes through the Formation of Stable C−H Bonds. Nano Lett 8(1):162–167. https://doi.org/10.1021/nl072325k
16. Chan S-P, Chen G, Gong XG, Liu Z-F (2001) Chemisorption of hydrogen molecules on carbon nanotubes under high pressure. Phys Rev Lett 87(20):205502
17. Lee SM, An KH, Lee YH, Seifert G, Frauenheim T (2001) A hydrogen storage mechanism in single-walled carbon nanotubes. J Am Chem Soc 123(21):5059–5063
18. Lee SM, Lee YH (2000) Hydrogen storage in single-walled carbon nanotubes. Appl Phys Lett 76(20):2877–2879
19. Ioannatos GE, Verykios XE (2010) H2 storage on single-and multi-walled carbon nanotubes. Int J Hydrogen Energy 35(2):622–628
20. Rather Su (2020) Preparation, characterization and hydrogen storage studies of carbon nanotubes and their composites: a review. Int J Hydrogen Energy 45(7):4653–4672. https://doi.org/10.1016/j.ijhydene.2019.12.055
21. Dinadayalane TC, Kaczmarek A, Łukaszewicz J, Leszczynski J (2007) Chemisorption of hydrogen atoms on the sidewalls of armchair single-walled carbon nanotubes. J Phys Chem C 111(20):7376–7383. https://doi.org/10.1021/jp066469j
22. Hod O, Barone V, Peralta JE, Scuseria GE (2007) Enhanced Half-Metallicity in Edge-Oxidized Zigzag Graphene Nanoribbons. Nano Lett 7(8):2295–2299. https://doi.org/10.1021/nl0708922
23. Ataca C, Aktürk E, Ciraci S (2009) Hydrogen storage of calcium atoms adsorbed on graphene: first-principles plane wave calculations. Phys Rev B 79(4):041406
24. Zhang J, Yan Y, Zong W, Li A, Qiao Z, Sun T (2017) Chemisorption of hydrogen on graphene: insights from atomistic simulations. J Phys: Condens Matter 29(19):195001. https://doi.org/10.1088/1361-648x/aa691a
25. Miwa RH, Martins TB, Fazzio A (2008) Hydrogen adsorption on boron doped graphene: anab initiostudy. Nanotechnology 19(15):155708. https://doi.org/10.1088/0957-4484/19/15/155708
26. Kim G, Jhi S-H (2009) Ca-Decorated Graphene-Based Three-Dimensional Structures for High-Capacity Hydrogen Storage. J Phys Chem C 113(47):20499–20503. https://doi.org/10.1021/jp907368t
27. Kumar D, Govindaraja T, Krishnamurty S, Kaliaperumal S, Pal S (2018) Dissociative chemisorption of hydrogen molecules on defective graphene-supported aluminium clusters: a computational study. Phys Chem Chem Phys 20(41):26506–26512. https://doi.org/10.1039/C8CP05711G
28. Hernández-Hernández A, Vallejo E, Martínez-Farías F, Pelayo JJ, Hernández-Hernández LA, Pescador-Rojas JA, Tamayo-Rivera L, Morales-Peñaloza A, López-Pérez PA, Cortes ER (2018) Changes to the dissociation barrier of H2 due to buckling induced by a chemisorbed hydrogen on a doped graphene surface. J Mol Model 24(9):244. https://doi.org/10.1007/s00894-018-3763-z
29. Sahaym U, Norton MG (2008) Advances in the application of nanotechnology in enabling a 'hydrogen economy.' J Mater Sci 43(16):5395–5429

30. Wang P, Wang AM, Wang YL, Zhang HF, Hu ZQ (2000) Decomposition behavior of MgH2 prepared by reaction ball-milling. Scripta Mater 43(1):83–87. https://doi.org/10.1016/S1359-6462(00)00370-5
31. Lee J, Farha OK, Roberts J, Scheidt KA, Nguyen ST, Hupp JT (2009) Metal–organic framework materials as catalysts. Chem Soc Rev 38(5):1450–1459
32. Li Y, Yang RT (2006) Significantly Enhanced Hydrogen Storage in Metal−Organic Frameworks via Spillover. J Am Chem Soc 128(3):726–727. https://doi.org/10.1021/ja056831s
33. Zlotea C, Campesi R, Cuevas F, Leroy E, Dibandjo P, Volkringer C, Loiseau T, Férey G, Latroche M (2010) Pd nanoparticles embedded into a metal-organic framework: synthesis, structural characteristics, and hydrogen sorption properties. J Am Chem Soc 132(9):2991–2997
34. Zhou C, Yao S, Wu J, Forrey RC, Chen L, Tachibana A, Cheng H (2008) Hydrogen dissociative chemisorption and desorption on saturated subnano palladium clusters (Pdn, n = 2–9). Phys Chem Chem Phys 10(35):5445–5451. https://doi.org/10.1039/B804877K
35. Malouche A, Oumellal Y, Ghimbeu CM, de Yuso AM, Zlotea C (2017) Exploring the hydrogen absorption into Pd–Ir nanoalloys supported on carbon. J Nanopart Res 19(8):270. https://doi.org/10.1007/s11051-017-3978-4
36. Matei Ghimbeu C, Puscasu A, Martinez de Yuso A, Zlotea C, Oumellal Y, Latroche M, Vix-Guterl C (2016) One-pot synthesis of tailored Pd–Co nanoalloy particles confined in mesoporous carbon. Microporous Mesoporous Mater 223:79–88. https://doi.org/10.1016/j.micromeso.2015.10.031
37. Adams BD, Ostrom CK, Chen S, Chen A (2010) High-performance Pd-based hydrogen spillover catalysts for hydrogen storage. J Phys Chem C 114(46):19875–19882
38. Campesi R, Cuevas F, Leroy E, Hirscher M, Gadiou R, Vix-Guterl C, Latroche M (2009) In situ synthesis and hydrogen storage properties of PdNi alloy nanoparticles in an ordered mesoporous carbon template. Microporous Mesoporous Mater 117(1–2):511–514
39. Kobayashi H, Morita H, Yamauchi M, Ikeda R, Kitagawa H, Kubota Y, Kato K, Takata M, Toh S, Matsumura S (2012) Nanosize-induced drastic drop in equilibrium hydrogen pressure for hydride formation and structural stabilization in Pd–Rh solid-solution alloys. J Am Chem Soc 134(30):12390–12393
40. Kobayashi H, Yamauchi M, Ikeda R, Kitagawa H (2009) Atomic-level Pd–Au alloying and controllable hydrogen-absorption properties in size-controlled nanoparticles synthesized by hydrogen reduction. Chem Commun 32:4806–4808
41. Psofogiannakis GM, Steriotis TA, Bourlinos AB, Kouvelos EP, Charalambopoulou GC, Stubos AK, Froudakis GE (2011) Enhanced hydrogen storage by spillover on metal-doped carbon foam: an experimental and computational study. Nanoscale 3(3):933–936

Chapter 9
Hydrogen Fuel Cells and Nanotechnology

9.1 Introduction

Fuel cells are regarded as the green power houses of the modern century. They are thought as small sources with the ability to turn hydrogen energy or economy into reality. The depletion of existing energy sources and increased pollution caused by their production and use are the major reasons behind the immense research in the area of hydrogen fuel cells [1, 2]. Fuel cells when run on pure hydrogen do not generate pollution and only produce clean heat and pure water as by-products. However, when mixture of hydrogen-rich reformate gas is used, harmful emissions (though less severe than fossil fuels) can be expected. The internal combustion engines, which utilized hydrogen as fuel, work as a mixture with air leading to extremely small pollution levels [3].

In simple terms, fuel cells are energy conversion tools. They are capable of converting chemical energy directly into electricity. The conversion of the energy does not involve any intermediate thermal or mechanical steps. Energy can be obtained whenever, and the fuel present in the cell is allowed to have the chemical reaction with the oxygen in air. In internal combustion engine, the combustion reaction takes place resulting in the formation of heat, a part of which is utilized in doing work, such as moving of the piston. In contrast to this, the fuel cells involve electrochemical reaction for the generation of energy. The energy is evolved as the mixture of heat and low-voltage direct current electrical energy. The electrical energy is involved in work, whereas the heat is wasted or can be employed for other purposes. In voltaic cells or galvanic cells, the chemical energy is converted into the electrical energy via electrochemical reactions. Any type of fuel cell is essentially a galvanic cell, and also it can be regarded as a battery. However, they are different from electroplater or electrolyzer which involves the consumption of electrical energy for driving a chemical reaction. A fundamental characteristic of fuel cells is that the usage of hydrogen and oxygen is determined by the electric current load [4].

S. Farrukh et al., *Nanotechnology and the Generation of Sustainable Hydrogen*, Green Energy and Technology, https://doi.org/10.1007/978-3-030-60402-8_9

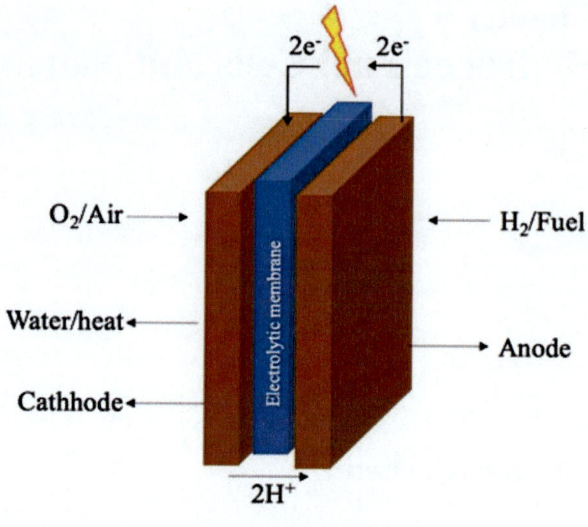

Fig. 9.1 Operation of hydrogen fuel cell

A fuel cell can work with wide range of fuels and oxidants. However, for practical purposes, hydrogen is regarded as the highly efficient fuel to be employed in fuel cells. This high efficiency of the hydrogen in fuel cells is associated with its extremely high electrochemical reactivity as compared to other fuels, like alcohols or hydrocarbons. The importance of hydrogen as the primary fuel of fuel cells can be acknowledged from the fact that the fuel cells, which do not directly utilized hydrogen as fuel, first convert other fuels into hydrogen. As for the oxidants are concerned, O_2 is the apparent choice owing to its abundance and remarkable reactivity [5]. A diagrammatic functioning of fuel cell is shown in Fig. 9.1.

A typical hydrogen fuel cell consists of two electrodes (cathode and anode) and an electrolyte membrane (mostly proton exchange membrane, i.e., PEM). The hydrogen and oxygen enter through the anode and cathode of a fuel cell, respectively [6]. At anode, the catalyst oxidized the hydrogen molecules and the obtained electrons move through the electric circuit, whereas the protons pass through the electrolytic membrane. The electron leads to the formation of electric current and heat, whereas the protons lead to the formation of water by combining with oxygen and electrons at cathode (reduction). When pure hydrogen is used as the fuel the chemical equations for the whole reaction can be given as [7].

$$\text{At anode:}\quad H_2 \rightarrow 2H^+ + 2e^- \qquad \text{(Oxidation)}$$

$$\text{At cathode:}\quad 1/2O_2 + 2H^+ + 2e^- \rightarrow H_2O \qquad \text{(Reduction)}$$

The hydrogen fuel cells utilized nanotechnology in a variety of ways. In the last few years, fuel cells have exhibited remarkable consistency and lower prices owing to the inclusion of nanomaterials in their production. The involvement of nanotechnology in the fabrication of fuels cells allows high aspect ratio, greater surface area which leads to more power generation, high energy densities, easy miniaturization, and longer shelf life. All of these characteristics are vital for the preparation of powerful fuel cell for transportable electric devices [8, 9]. This chapter will discuss various aspects of nanotechnology that are involved in hydrogen fuel cells.

9.2 Fuel Cell Electrolytic Membranes

Different types of nanomaterials have been employed in the hydrogen generation, decontamination, and storage in fuel cells [13]. In this passage, the nanostructured membranes employed in fuel cells will be discussed. Nanostructured membranes have garnered significant attention, particularly the nanostructured proton electrolyte membrane (PEM) of fuel cells, which have been extensively researched and studied [14]. The PEM is regarded as the "heart of the fuel cell." It is the membrane which has several important abilities like large proton conduction with electrical isolation, chemical and thermal stability, sufficient mechanical strength, and improved water administration features over widespread ranges of humidity and temperature [14]. The most commonly employed PEM in fuel cells is Nafion, which fundamentally consists of perfluorinated polymer [15]. PEM membranes have the issue of poor performance at high temperatures and low humidity [16]. Nanomaterials are being used for the eradication of this problem.

In an investigation, Nafion-silicon oxide-phosphotungstic acid nanocomposite membrane was fabricated via sol–gel method. The nanocomposite membranes were fabricated by casting the mixture of Nafion solution, phosphotungstic acid, and tetra ethoxy orthosilane solution. Infrared (IR) spectroscopic analysis confirms the peaks of silicon oxide and phosphotungstic acid and other inorganic and organic components of the nanocomposite membranes. When the membrane was employed in the fuel cell for analysis, it showed current density of 82 mA cm^{-2} at 0.6 V where pristine Nafion membrane depicted the current density of 30 m A cm^{-2} at 0.2 V at 40% relative humidity and 90 °C temperature. The internal resistance offered by the membranes was found to be in direct relation with the inorganic constituents. The prepared membranes performed better than commercial Nafion membranes in terms of internal resistance, Tafel constants, and overall performance [17].

In another investigation, modified Nafion nanocomposite membranes were fabricated for the purpose of achieving high proton conductivity and water preservation at lower humidity (~40%) and elevated temperatures (120 °C) along with enhanced mechanical and thermal strength of the membranes. The nanocomposite membranes were prepared by the integration of the metal oxide nanoparticles. Sol–gel method was used for the fabrication of Zr, Si, and Ti oxide nanoparticles and further these prepared nanoparticles were integrated in the pores of the Nafion membranes. The membranes fabricated with nanoparticles were entirely clear and homogeneous in composition, in contrast to the other membranes which turn cloudy due to casting with large size particles. All the modified nanocomposite membranes showed higher water retention capacity in comparison to the pristine Nafion membranes at 90 and 120 °C.

But the better proton conductivity than the Nafion membranes was only depicted by ZrO_2 sol–gel composite. This can be related to the higher acidity of zirconia-based sol–gel nanocomposite membranes as compared to the similar membranes of Ti and Si. The results of thermal gravimetric analysis (TGA) and dynamic mechanical analysis (DMA) showed superiority of nanocomposite membranes over pristine Nafion membraned in terms of glass transition temperature and decomposition [18].

Marianne et al. have employed SiO_2 nanoparticles for modification of Nafion 115 membranes. The prepared membranes were used to analyze the water sorption effect, effective mobility of proton, concentration of proton, and conductivity of proton. The membranes were prepared by sol–gel process. The silica nanoparticles were loaded in the range of 5.9–33.3 wt.% in the Nafion membranes. The density of the membranes was found to be in reverse relation with the silica content, and also the composite membranes depicted small dimensional changes with swelling in water, hence supporting the theory of rigid scaffolding formation within the membrane. The retention of the water increases with the increased loading of nanoparticles because of void formation in the membrane. Greater concentration of water in the membrane lowers the concentration of proton, hence reducing the conductivity of proton. Increased silica content resulted in decreased mobility of proton (this can be associated to enhanced tortuosity of the proton-conduction passage) also contributing to the reduction in conductivity of proton [19].

In another work, covalently interlinked nanocomposite hydroxide ion transporting membranes were fabricated on quaternized polysulfone and modified graphene oxide support. Fourier transform infrared attenuated total reflection spectroscopy (FTIR-ATR) and transmission electron microscopy (TEM) characterization were used to analyze the structure and morphology of the fabricated membranes. Different properties of the membrane like water sorption, swelling proportion, ionic conductivity, and mechanical strength were determined. The prepared nanocomposite membrane with functionalized graphene oxide content of 2% was found to be flexible and tough. The membrane depicted 19.44% water sorption and 1.27×10^{-2} S $cm^{-}1$ ionic conductivity at 60 °C and 14.90 MPa [20].

In an investigation, polybenzimidazole nanocomposite membranes doped with phosphoric acid and integrated with inorganic nanoparticles were fabricated. The polybenzimidazole polymer was integrated with varying amounts of different inorganic fillers like titanium dioxide, silicon dioxide, and zirconium phosphate (an inorganic proton conductor) followed by the doping of phosphoric acid. This resulted in the formation of high-temperature proton exchange membrane fuel cells. The fabricated membranes were analyzed for acid sorption and acid leaching abilities, mechanical strength, impedance analyses, and chemical stability in NN dimethylacetamide (DMAc). TGA characterization confirms enhanced thermal stability of the prepared membranes. The inorganic fillers were found to enhance the acid retention ability of the membranes. Electrochemical impedance spectroscopy (EIS) demonstrated that the integration of 5 wt. % ZrP or 5 wt. % SiO_2 enhances the conductivity of proton. However, the nanocomposite membrane integrated TiO_2 showed low conductivity values than pure polybenzimidazole membrane. The poor performance of the TiO_2 is associated to its non-uniform structure. The highest conductivity of proton was depicted by maximum phosphoric acid-doped ZrP-polybenzimidazole nanocomposite membrane. Nyquist plot analysis of membranes at varying temperatures revealed good agreement with Randel's circuit. The study concluded that these membranes are excellent choice for potential applications in high-temperature proton exchange membrane fuel cells [21].

9.3 Fuel Cell Electrodes

The fuel electrodes have also been benefitted by nanotechnology. A wide variety of fuel cells employ several kinds of electrodes for their operations. From the past several years, the nanotechnology has infiltrated this area of sustainable energy production.

A dual function of energy generation and storage is performed by a type of fuel cells, known as solid oxide regenerative fuel cells. They are capable of operating at high temperatures with minimum effects on environment, and hence an appropriate choice for energy managing units. The ability of the electrodes used in these fuel cells can be considerably enhanced by employing nanomaterials in place of bulk size materials. However, the use of nanomaterials for the purpose is difficult because of decrease in stability and control over the nanomaterial at elevated temperatures [22]. This problem has been addressed in a study. The study proposed the use of wet chemical infiltration technology which permits high-level controllability of nanoscale materials and their stability at high temperatures. In this study, researchers fabricated $Sm_{0.5}Sr_{0.5}CoO_3$ nanocatalyst electrodes for potential use in solid oxide regenerative fuel cells. The nanomaterials were fabricated via homogeneous precipitation. This was achieved by decomposition of urea in chemical solution, resulting in improved crystallization along with the efficient redistribution of precursor. The efficient precursor redistribution allows accurate adaptation of the phase purity and geometric properties. By adjusting the prime features of the nanocatalysts, an electrode can be produced that is very near to the ideal electrode model. The prepared electrodes were analyzed for their durability and performance in both electrolytic and fuel cells. The study proposed the aforementioned technology as reliable and cost-effective method for the production of high-temperature operating fuel cell electrodes [23].

$PrBaMn_2O_{5+\delta}$ is one of the excellent symmetrical solid oxide fuel cell electrode forming substances. *Yiheng* et al. reported the use of $PrBaMn_2O_{5+\delta}$ as backbone of both cathode and anode of symmetrical solid oxide fuel cells, and the electrodes were coated with Pr_6O_{11} nanocatalysts. The material was used as synergic catalysts to increase the efficiency of the electrodes. $PrBaMn_2O_{5+\delta}$ powder was synthesized via citric acid combustion procedure. For the purpose, the solution of $Pr(NO_3)_3 \cdot 6H_2O$, $Ba(NO_3)_2$, and $Mn(NO_3)_2 \cdot 4H_2O$ was made in distilled water. Ethylene glycol was mixed in the solution as the complexating agent. The pH of the solution was adjusted to ~8 by the addition of $NH_3.H_2O$. Afterward, the solution was stirred for 2 h, and then heated till the automatic combustion occurred. The product thus obtained was calcined at 950 °C for 4 h. The powder was then developed into electrodes by dry pressing method. For the coating of Pr_6O_{11} nanoparticles, the infiltration technique was used. In practice, the 1 M solution $Pr(NO_3)_3.6H_2O$ was prepared and then dripped onto the $PrBaMn_2O_{5+\delta}$ electrode backbone. The fabricated electrodes were characterized with hydrogen temperature-programmed reduction (H_2-TPR), field-emission scanning electron microscope (FESEM), X-ray diffraction (XRD), oxygen temperature-programmed desorption (O_2-TPD), X-ray photoelectron spectroscopy (XPS), and electrochemical measurements. The nanocatalyst-coated electrodes depicted decreased cathode and anode polarization resistance. The symmetrical cell power outputs were enhanced by ~75% at 800 °C by employing the nanocatalyst-coated electrodes. The electrodes were also found to be very stable in redox environment [24].

In an investigation, modified nanoflakes of cobalt enclosed with cobalt oxide were electrodeposited on different types of carbonaceous anodes. In this study, four forms of carbon anode, i.e., carbon paper, carbon cloth, graphite, and activated carbon were selected to be modified and then employed as high-performing anodes in microbial fuel cells. The characterizations results directed that modified cobalt nanoflakes that were enclosed by a thin cobalt oxide layer were developed on carbon anode surfaces. The cobalt oxide-covered nanoflakes of cobalt were deposited on the carbon electrodes via simple and efficient electrodeposition method, so as to eradicate the problem of the interfacial electron transfer and hydrophobicity of the electrodes. The deposition of thin layer of the nanoflakes on the electrodes appreciably improves the adhesion of microbes, the wettability of surface of the anode, and reduction in electron transfer resistance. The toxicity of the pristine cobalt is also reduced by the layer of cobalt oxide. When the modified electrodes were employed in the microbial fuel cells, they showed significant improvement in power generation. The modified carbon paper ele, carbon cloth, graphite, and activated carbon depicted 137, 103, 173, and 71% power generation, respectively. This proposed treatment technique represented a high performance, excellent microbial adhesion, easy fabrication, and scale-up anodes for MFC [25].

9.4 Fuel Cells and Nanocatalysts

The narrative of the fuel cells is incomplete without the involvement of the nanocatalyst. Several modern-day fuel cell technology researches take the route of nanocatalyst for the production of better function, efficient, and low-cost fuel cells. Some of the nanocatalyst utilized in a variety of fuel cells are discussed in Table 9.1.

Table 9.1 Nanocatalysts and their applications in various fuel cells

Nanocatalysts	Fuel cells	Property	References
Bimetallic zeolitic imidazolate frameworks on multiwalled carbon nanotubes	Polymer electrolyte membrane fuel cells	Improves conductivity and enhances reduction of oxygen	[26]
Carbon-nanotube-supported bioinspired nickel catalyst	Enzymatic fuel cell/proton exchange membrane fuel cell	Depicts reversible electrocatalytic activity for the $H_2/2H^+$	[27]
Nitrogen-doped graphene/Co–Ni alloy enclosed within carbon nanotubes	Microbial fuel cells	Shows remarkable catalytic activity for oxygen reduction	[28]
Pt/Co bimetallic nanotube	Proton exchange membrane fuel cell	Functions as cathode results in improved power density	[29]
Pt supported on graphene and carbon nanotube	Proton exchange membrane fuel cell	Shows catalytic activity for oxygen reduction	[30]
Ni–Fe-Se hydrogenases supported on functionalized carbon nanotubes	Hydrogen/air Enzymatic fuel cells	The structure of membrane allowed the efficient movement of oxygen-tolerant Ni–Fe-Se-hydrogenase	[31]

9.5 Conclusion

Nanomaterials are extensively involved in the fabrication and funtioning of nanomaterials. The fuel cells utillized nanomembranes, nanomaterial-based electrodes, and the nanocatalysts. The nanomaterials are reducing the size of the fuel cells and and improving their efficency day by day. Most of the naomaterials that are involved in fuel cells are based on nobel metals. The need of the hour is to develop the ecnomical alternatives of the nanomaterials that are involved in fuel cells, as many of the nanomaterials are based on nobel metals. The reduction in the cost and improvement in efficency of the fuel cells is absolute requirement for the hydrogen evolution.

References

1. St-Pierre J, Wilkinson DP (2001) Fuel cells: a new, efficient and cleaner power source. Am Instit Chem Eng AIChE J 47(7):1482
2. Trimm DL, Önsan ZI (2001) Onboard fuel conversion for hydrogen-fuel-cell-driven vehicles. Catal Rev 43(1–2):31–84

3. Granovskii M, Dincer I, Rosen MA (2006) Economic and environmental comparison of conventional, hybrid, electric and hydrogen fuel cell vehicles. J Power Sources 159(2):1186–1193
4. Hsiung S (2001) Hydrogen fuel cell engines and related technologies
5. Spiegel C (2007) Designing and building fuel cells, vol 87. Citeseer
6. Haile SM (2003) Fuel cell materials and components. Acta Mater 51(19):5981–6000
7. Cassidy M, Ouweltjes JP, Dekker N (2010) Going beyond hydrogen: Non-hydrogen fuels, re-oxidation and impurity effects on solid oxide fuel cell anodes. Rsc Energy Environ S:153–189
8. Paul DR, Robeson LM (2008) Polymer nanotechnology: nanocomposites. Polymer 49(15):3187–3204
9. Zäch M, Hägglund C, Chakarov D, Kasemo B (2006) Nanoscience and nanotechnology for advanced energy systems. Curr Opin Solid State Mater Sci 10(3–4):132–143
10. Thiam HS, Daud WRW, Kamarudin SK, Mohammad AB, Kadhum AAH, Loh KS, Majlan EH (2011) Overview on nanostructured membrane in fuel cell applications. Int J Hydrogen Energy 36(4):3187–3205. https://doi.org/10.1016/j.ijhydene.2010.11.062
11. Jiang SP (2012) Nanoscale and nano-structured electrodes of solid oxide fuel cells by infiltration: advances and challenges. Int J Hydrogen Energy 37(1):449–470
12. Rajalakshmi N, Lakshmi N, Dhathathreyan KS (2008) Nano titanium oxide catalyst support for proton exchange membrane fuel cells. Int J Hydrogen Energy 33(24):7521–7526
13. Tegart G (2009) Energy and nanotechnologies: Priority areas for Australia's future. Technol Forecast Soc Chang 76(9):1240–1246
14. Wannek C, Glüsen A, Stolten D (2010) Materials, manufacturing technology and costs of fuel cell membranes. Desalination 250(3):1038–1041
15. Elfring GJ, Struchtrup H (2007) Thermodynamic considerations on the stability of water in Nafion. J Membr Sci 297(1–2):190–198
16. Inaba M, Kinumoto T, Kiriake M, Umebayashi R, Tasaka A, Ogumi Z (2006) Gas crossover and membrane degradation in polymer electrolyte fuel cells. Electrochim Acta 51(26):5746–5753
17. Mahreni A, Mohamad AB, Kadhum AAH, Daud WRW, Iyuke SE (2009) Nafion/silicon oxide/phosphotungstic acid nanocomposite membrane with enhanced proton conductivity. J Membr Sci 327(1):32–40. https://doi.org/10.1016/j.memsci.2008.10.048
18. Jalani NH, Dunn K, Datta R (2005) Synthesis and characterization of Nafion®-MO2 (M=Zr, Si, Ti) nanocomposite membranes for higher temperature PEM fuel cells. Electrochim Acta 51(3):553–560. https://doi.org/10.1016/j.electacta.2005.05.016
19. Rodgers MP, Shi Z, Holdcroft S (2008) Transport properties of composite membranes containing silicon dioxide and Nafion®. J Membr Sci 325(1):346–356. https://doi.org/10.1016/j.memsci.2008.07.045
20. Liu J, Qu R, Peng P, Liu W, Chen D, Zhang H, Liu X (2016) Covalently functionalized graphene oxide and quaternized polysulfone nanocomposite membranes for fuel cells. RSC Adv 6(75):71305–71310. https://doi.org/10.1039/C6RA12822J
21. Özdemir Y, Üregen N, Devrim Y (2017) Polybenzimidazole based nanocomposite membranes with enhanced proton conductivity for high temperature PEM fuel cells. Int J Hydrogen Energy 42(4):2648–2657. https://doi.org/10.1016/j.ijhydene.2016.04.132
22. Yokokawa H (2003) Understanding materials compatibility. Annu Rev Mater Res 33(1):581–610
23. Joong Yoon K, Biswas M, Kim H-J, Park M, Hong J, Kim H, Son J-W, Lee J-H, Kim B-K, Lee H-W (2017) Nano-tailoring of infiltrated catalysts for high-temperature solid oxide regenerative fuel cells. Nano Energy 36:9–20. https://doi.org/10.1016/j.nanoen.2017.04.024
24. Gu Y, Zhang Y, Zheng Y, Chen H, Ge L, Guo L (2019) PrBaMn2O5+δ with praseodymium oxide nano-catalyst as electrode for symmetrical solid oxide fuel cells. Appl Catal B 257:117868. https://doi.org/10.1016/j.apcatb.2019.117868
25. Mohamed HO, Abdelkareem MA, Obaid M, Chae S-H, Park M, Kim HY, Barakat NAM (2017) Cobalt oxides-sheathed cobalt nano flakes to improve surface properties of carbonaceous electrodes utilized in microbial fuel cells. Chem Eng J 326:497–506. https://doi.org/10.1016/j.cej.2017.05.166

26. Zhang C, Wang Y-C, An B, Huang R, Wang C, Zhou Z, Lin W (2017) Networking pyrolyzed zeolitic imidazolate frameworks by carbon nanotubes improves conductivity and enhances oxygen-reduction performance in polymer-electrolyte-membrane fuel cells. Adv Mater 29(4):1604556. https://doi.org/10.1002/adma.201604556
27. Gentil S, Lalaoui N, Dutta A, Nedellec Y, Cosnier S, Shaw WJ, Artero V, Le Goff A (2017) Carbon-nanotube-supported bio-inspired nickel catalyst and its integration in hybrid hydrogen/air fuel cells. Angew Chem Int Ed 56(7):1845–1849. https://doi.org/10.1002/anie.201611532
28. Hou Y, Yuan H, Wen Z, Cui S, Guo X, He Z, Chen J (2016) Nitrogen-doped graphene/CoNi alloy encased within bamboo-like carbon nanotube hybrids as cathode catalysts in microbial fuel cells. J Power Sources 307:561–568. https://doi.org/10.1016/j.jpowsour.2016.01.018
29. Zeng Y, Shao Z, Zhang H, Wang Z, Hong S, Yu H, Yi B (2017) Nanostructured ultrathin catalyst layer based on open-walled PtCo bimetallic nanotube arrays for proton exchange membrane fuel cells. Nano Energy 34:344–355. https://doi.org/10.1016/j.nanoen.2017.02.038
30. Remy E, Thomas Y, Guetaz L, Fouda-Onana F, Jacques P-A, Heitzmann M (2018) Optimization and tunability of 2D graphene and 1D carbon nanotube electrocatalysts structure for PEM fuel cells. Catalysts 8(9):377
31. Gentil S, Che Mansor SM, Jamet H, Cosnier S, Cavazza C, Le Goff A (2018) Oriented immobilization of [NiFeSe] hydrogenases on covalently and noncovalently functionalized carbon nanotubes for H2/Air enzymatic fuel cells. ACS Catal 8(5):3957–3964. https://doi.org/10.1021/acscatal.8b00708

Chapter 10
Hydrogen Future: Toward Industrial Applications

Nowadays the high levels of pollution, contaminated air, change in climate, accidents, obstruction, land usage, and the noise pollution declared modern transportation as "unsustainable." In the growing metro cities across the world, the air pollution from transportation is becoming serious threat to the human existence. The emission of greenhouse gases from automobiles and burning of fossil fuel are menacing to mankind, along with the serious decline in fossil fuel reserves. In the coming years, a continuous increase in human population is estimated, which will enhance global fuel demand [1]. Other than this enhanced consumption of fuel leads to increased emission of carbon dioxide and overall temperature of the earth [2].

The use of hydrogen as the fuel in different areas such as industry and transportation is an effective measure against increasing pollution and depleting fuel sources. However, at present, the generation, storage, and use of hydrogen are not economical. The factors which can affect the hydrogen fuel supply are demand level, availability of resource, geographic features, and most importantly the technology development for hydrogen generation and storage [3]. There are several elements which should be considered for the supply of hydrogen fuel as the alternative to commonly employed fuels. These elements are applications for which fuel is required, generation processes of the fuel, the possible pathways, and the problems linked to the shift toward large-scale development.

Currently, there are some technical barriers in the path of transition from carbon-based energy economy to a hydrogen-based system. The first and foremost is the cost reduction in hydrogen energy, and the production, storage, and delivery. The overall prices of hydrogen fuel and hydrogen-based system must be controlled and expensive items must be replaced with more efficient and economical alternatives. The cost of fuel cell and other hydrogen-based systems must also be reduced [4]. Otherwise, it is near impossible to make hydrogen a commercial alternative to the carbon-based fuel. Secondly, the hydrogen storage mediums are not enough for both automobile and stationary (battery) applications. The new and better technology must be developed with the eradication of shortcomings in previous systems. The

S. Farrukh et al., *Nanotechnology and the Generation of Sustainable Hydrogen*, Green Energy and Technology, https://doi.org/10.1007/978-3-030-60402-8_10

principle future markets for hydrogen are predominantly based on following factors [5]:

i. The price of hydrogen fuel.
ii. The relative rate of advancements and developments in different technologies associated with hydrogen.
iii. Potential constraints on generation or emission of greenhouse gases
iv. The price of contending energy structures.

Hydrogen has scored the prime position in the race of future fuels, owing to its several economic, social, and environmental advantages. It has a long-lasting ability to minimize the dependency on fossil oil (which has to be imported at high price in several cases) and minimizing the emission of carbon oxides and other greenhouse gases [6].

Many areas in the generation of hydrogen are assumed to be improved with the use of nanotechnology in terms of cost and efficiency. The abovementioned assumption has been drawn from the ongoing research on nanotechnology, in production and storage of hydrogen. In the previous chapters, the authors have tried to explain the importance of nanotechnology in the generation and storage of sustainable hydrogen. Though nanotechnology is paving the path for the smooth supply of hydrogen to the consumers for a variety of applications. However, the room for progress and development is still there for the commercialization of the fuel.

Here some of the potential solutions are listed, which nanotechnology can offer toward the commercialization of hydrogen:

1. One of the most important sources of hydrogen generation is water. Water liberates hydrogen and oxygen via water splitting reaction. The hydrogen yield from water splitting reactions has been augmented to several folds by the use of different nanomaterials. For instance, quantum dots of carbon supported on monolayer of layer C_3N have increased the efficiency of water splitting reaction without the mixing of the gases [7]. In similar manner, various metallic nanoparticles [8], nanotubes [9], nanosheets [10], and other nanomaterials are serving the splitting of water for generation of hydrogen. The nanomaterials are better alternatives to the conventional catalysts of the reactions. The current study suggests that the nanomaterials must be optimized on the basis of required application, geographic location, and the available materials for achieving the efficient water splitting reaction.
2. Hydrocarbons are also used as the substrate for the generation of hydrogen. Different processes of hydrocarbon decarbonization for H_2 generation utilize nanotechnology. For instance, steam reforming of hydrocarbons for hydrogen production is proving to be more resourceful via the use of nanotechnology [11]. In similar way, the partial oxidation [12] and photocatalytic activation [13] of hydrocarbons for generation of hydrogen is proven to be benefitted from nanotechnology. The current investigation proposed the utilization of less used or harmful hydrocarbons (aromatics) for the production of hydrogen instead of highly used hydrocarbons (methane). The applications of nanotechnology

should be investigated for hydrogen generation from harmful or less useful hydrocarbons.

3. Several investigators have researched the generation of hydrogen from hydrogen sulfide, assisted with transition metal-based nanomaterials. The nanomaterials based on Cd [14], Zn [15], and Ti [16] are proven to be very useful in decomposition of hydrogen sulfide for liberation of hydrogen. The authors proposed that the capturing of H_2S should also be studied with nanomaterials along with the decomposition. The H_2S capturing is the major obstacle in the path of hydrogen generation from the substance.
4. Storage of hydrogen via physisorption and chemisorption is considerably improved with the utilization of nanotechnology. Different nanomaterials like fullerenes [17], graphene [18], and nanotubes have played an important role in this regard [19].
5. Nanotechnology is currently being used for the optimization of materials and methods, employed in fuel cell development for superior functioning, stability, and storage of hydrogen [20].
6. The enhancement in catalytic activity of a substance at nanoscale is an established fact owing to the small size and enhanced surface area of the nanoparticles. In a comparative study, the researchers at Technical University of Munich have utilized nanoparticles of Pt and have achieved double catalytic performance in comparison to the currently available materials in fuel cell applications. The researchers have found that the cluster of only 40 Pt atoms is sufficient for producing equivalent results as exhibited by mass of material, hence leading to considerable cost reductions (as Pt is very expensive) [21]. The researchers suggested the use of economical cobalt–graphene catalyst as the replacement of expensive Pt materials in fuel cells [22].
7. There is radical improvement in the performance of the fuel cells by using the nanomembranes. Several studies have claimed that nanomembranes are better option as compared to the conventional membranes for fuel cell applications [23].

In short, the nanotechnology is actively providing useful services in the commercialization of hydrogen. In this work, the authors tried to represent the current picture of nanotechnology in generation and storage of hydrogen. The aim of the work is to aid the researchers and scientists to explore new dimensions in the field of sustainable energy based on nanotechnology. In this way, the researchers can follow the path of nanotechnology to overcome shortcomings and problems, hindering the way to clean and sustainable hydrogen.

References

1. Ball M, Wietschel M (2009) The future of hydrogen–opportunities and challenges. Int J Hydrogen Energy 34(2):615–627
2. Watson RT, Albritton DL, Dokken DJ (2001) Climate change 2001: synthesis report. Cambridge University Press Cambridge, UK
3. Sharma S, Ghoshal SK (2015) Hydrogen the future transportation fuel: From production to applications. Renew Sustain Energy Rev 43:1151–1158. https://doi.org/10.1016/j.rser.2014.11.093
4. Edwards PP, Kuznetsov VL, David WIF, Brandon NP (2008) Hydrogen and fuel cells: towards a sustainable energy future. Energy Policy 36(12):4356–4362
5. Forsberg CW (2005) Hydrogen markets: implications for hydrogen production technologies. Int J Hydrogen Energy
6. Balat M (2008) Potential importance of hydrogen as a future solution to environmental and transportation problems. Int J Hydrogen Energy 33(15):4013–4029
7. Wang X, Jiang X, Sharman E, Yang L, Li X, Zhang G, Zhao J, Luo Y, Jiang J (2019) Isolating hydrogen from oxygen in photocatalytic water splitting with a carbon-quantum-dot/carbon-nitride hybrid. J Mater Chem A 7(11):6143–6148
8. Wu W, Wu Y, Zheng D, Wang K, Tang Z (2019) Ni@ Ru core-shell nanoparticles on flower-like carbon nanosheets for hydrogen evolution reaction at All-pH values, oxygen evolution reaction and overall water splitting in alkaline solution. Electrochim Acta 320:134568
9. Park JH, Kim S, Bard AJ (2006) Novel carbon-doped TiO2 nanotube arrays with high aspect ratios for efficient solar water splitting. Nano Lett 6(1):24–28
10. Zhou X, Gao Q, Li X, Liu Y, Zhang S, Fang Y, Li J (2015) Ultra-thin SiC layer covered graphene nanosheets as advanced photocatalysts for hydrogen evolution. J Mater Chem A 3(20):10999–11005
11. Bej B, Pradhan NC, Neogi S (2013) Production of hydrogen by steam reforming of methane over alumina supported nano-NiO/SiO2 catalyst. Catal Today 207:28–35. https://doi.org/10.1016/j.cattod.2012.04.011
12. Bkour Q, Zhao K, Scudiero L, Han DJ, Yoon CW, Marin-Flores OG, Norton MG, Ha S (2017) Synthesis and performance of ceria-zirconia supported Ni-Mo nanoparticles for partial oxidation of isooctane. Appl Catal B 212:97–105. https://doi.org/10.1016/j.apcatb.2017.04.055
13. Kuo H-L, Kuo C-Y, Liu C-H, Chao J-H, Lin C-H (2007) A highly active bi-crystalline photocatalyst consisting of TiO2 (B) nanotube and anatase particle for producing H2 gas from neat ethanol. Catal Lett 113(1):7–12. https://doi.org/10.1007/s10562-006-9009-1
14. Kanade KG, Baeg J-O, Mulik UP, Amalnerkar DP, Kale BB (2006) Nano-CdS by polymer-inorganic solid-state reaction: Visible light pristine photocatalyst for hydrogen generation. Mater Res Bull 41(12):2219–2225. https://doi.org/10.1016/j.materresbull.2006.04.031
15. Chaudhari NS, Bhirud AP, Sonawane RS, Nikam LK, Warule SS, Rane VH, Kale BB (2011) Ecofriendly hydrogen production from abundant hydrogen sulfide using solar light-driven hierarchical nanostructured ZnIn2S4 photocatalyst. Green Chem 13(9):2500–2506
16. Chaudhari NS, Warule SS, Dhanmane SA, Kulkarni MV, Valant M, Kale BB (2013) Nanostructured N-doped TiO 2 marigold flowers for an efficient solar hydrogen production from H_2S. Nanoscale 5(19):9383–9390
17. Mert H, Deniz CU, Baykasoglu C (2020) Monte Carlo simulations of hydrogen adsorption in fullerene pillared graphene nanocomposites. Mol Simul 46(8):650–659. https://doi.org/10.1080/08927022.2020.1758696
18. Kosasih EA, Kurniawan B, Zulkarnain IA (2016) Optimization of hydrogen storage capacity by physical adsorption on open-ended single-walled carbon nanotube as diameter function. Int J Technol 7(2):264–273
19. Cabria I, López MJ, Alonso JA (2006) Density functional calculations of hydrogen adsorption on boron nanotubes and boron sheets. Nanotechnology 17(3):778
20. Abdalla AM, Hossain S, Azad AT, Petra PMI, Begum F, Eriksson SG, Azad AK (2018) Nanomaterials for solid oxide fuel cells: a review. Renew Sustain Energy Rev 82:353–368

21. Garlyyev B, Kratzl K, Rück M, Michalička J, Fichtner J, Macak JM, Kratky T, Günther S, Cokoja M, Bandarenka AS, Gagliardi A, Fischer RA (2019) Optimizing the Size of Platinum Nanoparticles for Enhanced Mass Activity in the Electrochemical Oxygen Reduction Reaction. Angew Chem Int Ed 58(28):9596–9600. https://doi.org/10.1002/anie.201904492
22. Guo S, Zhang S, Wu L, Sun S (2012) Co/CoO Nanoparticles Assembled on Graphene for Electrochemical Reduction of Oxygen. Angew Chem Int Ed 51(47):11770–11773. https://doi.org/10.1002/anie.201206152
23. Labrador NY, Songcuan EL, De Silva C, Chen H, Kurdziel SJ, Ramachandran RK, Detavernier C, Esposito DV (2018) Hydrogen evolution at the buried interface between Pt thin films and silicon oxide nanomembranes. ACS Catal 8(3):1767–1778